Science and Fiction

Science and Fiction – A Springer Series

This collection of entertaining and thought-provoking books will appeal equally to science buffs, scientists and science-fiction fans. It was born out of the recognition that scientific discovery and the creation of plausible fictional scenarios are often two sides of the same coin. Each relies on an understanding of the way the world works, coupled with the imaginative ability to invent new or alternative explanations—and even other worlds. Authored by practicing scientists as well as writers of hard science fiction, these books explore and exploit the borderlands between accepted science and its fictional counterpart. Uncovering mutual influences, promoting fruitful interaction, narrating and analyzing fictional scenarios, together they serve as a reaction vessel for inspired new ideas in science, technology, and beyond.

Whether fiction, fact, or forever undecidable: the Springer Series "Science and Fiction" intends to go where no one has gone before!

Its largely non-technical books take several different approaches. Journey with their authors as they

- Indulge in science speculation—describing intriguing, plausible yet unproven ideas;
- Exploit science fiction for educational purposes and as a means of promoting critical thinking;
- Explore the interplay of science and science fiction—throughout the history of the genre and looking ahead;
- Delve into related topics including, but not limited to: science as a creative process, the limits of science, interplay of literature and knowledge;
- Tell fictional short stories built around well-defined scientific ideas, with a supplement summarizing the science underlying the plot.

Readers can look forward to a broad range of topics, as intriguing as they are important. Here just a few by way of illustration:

- Time travel, superluminal travel, wormholes, teleportation
- Extraterrestrial intelligence and alien civilizations
- Artificial intelligence, planetary brains, the universe as a computer, simulated worlds
- Non-anthropocentric viewpoints
- Synthetic biology, genetic engineering, developing nanotechnologies
- Eco/infrastructure/meteorite-impact disaster scenarios
- Future scenarios, transhumanism, posthumanism, intelligence explosion
- Virtual worlds, cyberspace dramas
- Consciousness and mind manipulation

More information about this series at http://www.springer.com/series/11657

V. Anne Smith

A Code for Carolyn

A Genomic Thriller

 Springer

V. Anne Smith
School of Biology
University of St. Andrews
St. Andrews, UK

ISSN 2197-1188   ISSN 2197-1196   (electronic)
Science and Fiction
ISBN 978-3-030-04551-7   ISBN 978-3-030-04553-1   (eBook)
https://doi.org/10.1007/978-3-030-04553-1

Library of Congress Control Number: 2018964419

This Springer imprint is published by the registered company Springer Nature Switzerland AG
The registered company address is: Gewerbestrasse 11, 6330 Cham, Switzerland

Contents

Part I

The Novel

1

A Code for Carolyn: A Genomic Thriller

Chapter One

Carolyn looked out over the lecture theatre, a sea of bent heads facing her as students scribbled notes on tablets or docfilm. She lifted a docfilm notebook and shook it. "Imagine this notebook is your DNA. Each time a new copy is made, the ends get just a little bit shorter." She tore a page of the grey docfilm out from the front and back. This analogy had worked better the way she had seen it in her student days, when people wrote on paper instead of docfilm. She scrubbed the top left corners of the sheets with a magnet, covering the time with some extra patter. "Remember, this is because DNA polymerase can't copy that last little bit at the linear end, under the terminal RNA primer." Finally, the pages went pink: disconnected. She lifted the pair.

The students gasped. Although perhaps it worked just as well now. Even though docfilm was backed up on the cloud, recovering data from disconnected film was not cheap.

She pointed at a young man in the third row, staring at her with his mouth parted. "If this was your DNA, where would you want your genes?"

"In the middle!" He clutched his docfilm pad to his chest, as if protecting it from his suddenly irrationally destructive lecturer.

She grinned. "Exactly. That's what your cells do. The telomeres are like pages of nonsense on either side of the genes, protecting their valuable information from being eaten away." A soft chime from the lectern's terminal sounded. One minute left. "But that can't go on forever, otherwise after a couple generations the protection would be gone. Next time we'll cover how

© Springer Nature Switzerland AG 2019
V. A. Smith, *A Code for Carolyn*, Science and Fiction,
https://doi.org/10.1007/978-3-030-04553-1_1

eukaryotic organisms keep their DNA from vanishing as our cells divide." She flicked to the next slide, leaving the word *Telomerase* on the screen as a teaser.

She logged out from the University's terminal, lifted her coat and the remains of the notebook, and turned to the queue that had formed, apparently instantaneously, in front of her.

"Dr Gray, can I ask a question about the essay?" The woman in front hugged a docfilm notebook, eyes flicking occasionally to the pink sheets peeking out from Carolyn's.

Carolyn suppressed her amusement. Today's students treated docfilm with the same reverence Carolyn had developed for printed books, possibly more. An institutional license meant she could pipe data between any registered sheets and her desktop terminal. Individuals did not have that luxury. She smiled. "Certainly."

They were always so nervous about big assignments; she remembered herself at that stage and did what she could to reassure them. And as the age difference between her own newly-teenage daughter and her students grew steadily less, the students somehow seemed younger and more vulnerable. Twelve questions about the essay later, Carolyn finished buttoning her coat and followed the last students out of the theatre.

The London air was wintery chill, a change which had happened only this week. She pulled up her collar to block the cold air buffeting her cheeks and bent into the wind funnelled between University buildings. Tuning out her surroundings, she planned the rest of her day. First stop would be the lab, to check on Frank's transformations. He had been stalled for weeks. If the last protocol changes had not helped, he needed to redesign his construct. PhD projects had a tight time frame. Then there was Tonya's paper to read, which the postdoc had sent three days ago. Plus two grant referee reports and several recommendation letters for undergraduates. Also … She reached the door to her building before she had finished her mental review. It did not slide open at her approach.

Surprised, she stared at the reflection of her tall shadowed form in the glass for a moment, her thick head of curls blowing slightly in the wind, before loosening her coat to get at her wallet. The doors were normally unlocked during working hours. She waved her wallet past the sensor. The doors parted.

A uniformed man stood inside. A patch over his left breast showed a black and white checkerboard stripe over a stylised graduation cap and the words *London Science University* in an arc underneath: campus police. "ID?"

She had just used it to get in. What was going on? She retrieved her wallet a second time and pulled out her ID card.

"Thank you, Dr Gray."

"You're welcome," she replied automatically and walked past. She looked back. He stood in the same position, staring at the closed door, with his hand raised to his face. Reporting her entrance? She pushed aside her curiosity. She had enough to do otherwise.

She turned down the next corridor. More uniformed people milled about, concentrating around the second door on the right: her lab! She strode forward, heart thumping. "What's going on?"

"Dr Gray?" A short blonde woman in a suit blocked her path. Her hair was pulled into a high ponytail, and a tiny pin on her lapel echoed the shields of the uniformed campus police. "I'm afraid there's been an incident. I'm Detective Roberts, and I'll be handling your case."

What accident would require all these police? Fire? It would smell more. Acid? She stepped sideways around the woman and made it to the taped-off door of her lab. The room looked like a small tornado had been through. Explosion?

But nothing was broken. Bottles stood upright, just off the shelves. Cabinets were open. Docfilms lay in great grey slicks on lab benches and the floor. An e-lab notebook sat on the bench to her right, emitting intermittent static as its screen alternately distorted and restored. She could not make sense of what she was looking at. "Where are my lab members?"

"In the coffee room." Detective Roberts appeared beside her. "Let's speak there. We have some questions."

I have some questions, thought Carolyn. But she followed the detective down the corridor. Two of her PhD students and a postdoc sat on the faded blue couches. Frank hunched over, cupping a dark, steaming drink to his chest. Carolyn could smell the sweetness of hot chocolate from the doorway. Aya leaned back next to him, also with steaming hot chocolate. Her long dark hair was loose; she had not yet been into the lab. The postdoc, Mohan, sat hunched like Frank, but without the drink, his fingers steepled.

"Have a seat," said Detective Roberts. Carolyn resisted an impulse to remind the detective that *she* was the intruder and sat beside Mohan. "Forensics will be done before lunch. You can enter the lab then."

"Is anything missing?" asked Aya.

"You'll have to tell us." Detective Roberts pulled out a docfilm pad and stylus. She tapped it a few times. "We've already been over this morning's events," she said, likely for Carolyn's benefit. "Frank arrived first, entered the lab to check his transforms, then—"

"Transformations," Frank said, correcting her.

"Did they work?" Carolyn asked. The lab's state was a shock, but those transformations were key to his project.

"Yes!" Frank grinned. "Some. Half the colonies were really small. I'd like you to look."

"Exactly half?" That was curious. "Were they petites?" Petites resulted from mitochondrial dysfunction. But the construct should not be interfering with the mitochondria yet. "I—"

"Ahem." The detective tapped her docfilm.

Carolyn's cheeks heated. For a moment, she had forgotten the circumstances.

"Then Frank phoned Aya, who phoned Mohan, who first checked his assays …"—Detective Roberts paused, as if expecting a correction, but Mohan nodded—"… then finally called us."

"What did you find?" Carolyn asked.

"Not much more than you saw. The lab has been disturbed, clearly, but not destructively. HazMat's come and gone; your chemicals are fine. Nothing's broken. It's just all been gone through."

That matched the brief look Carolyn had had. "Who could have done this?"

"That's what we want to ask you. Do you have any competitors? Secrets that someone might want?"

Carolyn shook her head. There were people in the same field, but they were collaborators half the time.

"But you have industry funding?"

"Of course." Governments did not fund science anymore. They stuck to lower-budget arts and humanities, and activities where the public could enjoy the fruits of their tax-funded effort directly. After basic science had nearly died out half a century ago in the quest for 'impact', it was, ironically, corporations who now had both the bankroll and the long view that enabled them to fund research without immediate application.

"Could a funder's competitor be trying to steal your work?" The detective absently flipped her stylus over her fingers.

"I have free disclosure contracts." Industry did not bother with IP issues for basic research like Carolyn's; they had more lucrative projects on which to concentrate their efforts. Plus, it left the public with the illusion that little had changed, even though corporations now owned most of the information in the world.

"Have you had any run-ins with student groups? Anticorporate ones, or perhaps some harsh marking?"

"You think students might have done this?" Her heart dropped at the thought. Not her students!

Detective Roberts shrugged. "It has hallmarks of a student prank, and they would have easiest access to the building." She frowned and tapped her stylus against her pad.

Carolyn realised she had not answered the question. "No, no run-ins." Few students in her intro biology class would be anticorporate, at least openly, if they wanted to work in science. "Our marks are all overseen; my module is pretty average." The students were nervous about the essay, but eager-nervous, not angry-nervous.

Detective Roberts handed across her card. "We'll be in touch, but call if you get any sudden ideas." She stood.

Carolyn tapped the card against the multicard in her wallet. *Roberts, Susan Detective* flashed briefly, then faded. She offered the small piece of docfilm back to the detective. "That's it?" Carolyn was not sure what more she expected. Answers, perhaps?

"Not much we can do until forensics is done. They'll check your elly-books for downloads. That might key us in to whether we're looking at a thief or a prank."

After about an hour of nervous speculation, they were let into the lab. Carolyn helped put things back in order. Despite the chaos, no projects had been disrupted. Her heart shied away from such an act being a student prank, although the alternative was perhaps more distressing. Who could want what from her lab?

Nothing obvious was missing, but with the proliferation of random solutions and a refrigerator full of cling film-wrapped plates, it was hard to tell. Yet nothing should be of any import. Any half-built constructs would do no one any good, not unless they were also studying mechanisms of mitochondrial turnover and wanted partially built tools. Her science was decidedly of the basic kind, unlike her parents' had been.

She had deliberately chosen research directions different as possible from that of her deceased, infamous parents. Somehow, that teenage promise to stay away from science, and biology in particular, had not stuck. But she had stayed as far away from synthetic organisms as modern biotechnology allowed.

She contemplated a stack of plates containing synthetically modified *S. cerevisiae*—not that far indeed. The date on the plates was two years ago. She dumped them in an autoclave bag.

* * *

It was mid-afternoon by the time Carolyn reached her office, and she had not yet eaten lunch. She collapsed into her chair, her mind on the lab. The possibility that students would target her hurt. She enjoyed teaching the younger generation, sharing her fascination with the biological world. But if it was an anticorporate group, it would not be personal. They thought all scientists had sold out, taking industry funding. Why *her* lab, though?

She should report the incident to her funders. She spun in her chair to pull out the file drawer where she kept her contracts. Instead of her tidy files, it held a docfilm slick like in the lab.

Her stomach dropped. They had been here too.

She jerked open the remaining drawers. Every one was disordered, though less so on the upper left. Whoever had done this must have started there and gotten messier as they went. And, she supposed, they had paid no heed at all when they reached the lab.

At least it was unlikely to be a student prank, then, her analytical mind told her, while her anxiety skyrocketed. She must have taken some comfort in the idea that it was an easily explainable prank, no matter how much it hurt. The thought that some strangers had ransacked her lab for an unknown purpose was like seeing the destruction all over again.

Just call the detective, she told herself. She counted three breaths. The detective was a professional. She would know what to do.

Carolyn turned back to her desk and pulled out her multicard. Something pale underneath the phone caught her eye. It was an envelope, off-white, made of paper. She crinkled it. Paper inside as well. Whatever was written there was not backed up on the cloud. The oddity of it pushed past her anxiety as curiosity surged.

She flipped the envelope over and squinted at the cursive writing on the front. Who wrote in cursive anymore? She read, *Carolyn Schwarz*.

Her breath stopped. No one knew that name!

Well, her uncle and his family did. Plus everyone she had attended high school with. But since then—she had changed her surname to that of her uncle's after the discovery of her parents' deception. She wanted never again to be connected to Baby C of geneticists Dr and Dr Schwarz.

Unbidden, memories of the news headlines arose: *'First' Synthesised Human … NOT, Human Hoax*, and others of the same ilk. Her parents had not, it turned out, made genomics history two decades ahead of its time by creating her DNA in a lab. She never learned who leaked the results of her high-school genetics exercise, where her karyotype revealed itself—persistently, through more than a dozen repetitions in her disbelief—to have three X chromosomes. No synthesised genome would have such obvious traces of biological origin:

an extra chromosome from an error during the formation of her mother's egg. She *was* a genetic anomaly, just not the scientific breakthrough they had claimed.

At least the headlines had only been accompanied by the same infant shots as the initial claim. Her parents' former employer, Vivcor, had stepped in to limit the damage. They had helped her change her name as well; it was their reputation that had been impacted, and they were happy that she wanted to make her hoax identity vanish. She had not intended to go into science, especially biology—the opposite, in fact. But perhaps her parents had given her more than her slightly defective genes, for biology drew her in and kept her. That her colleagues might discover her secret terrified her. Would they trust her science, knowing she was the offspring—the hoax itself—of the most infamous case of scientific misconduct this century? Baby C was on its way to joining the Piltdown Man in textbooks.

She opened the envelope with sweaty hands. Inside, written in the same cursive, was, *Where is it?* The words started neat, then scrawled, echoing the disruption to her file cabinets. She flipped the paper over then opened the envelope further, feeling for anything else.

Annoyance replaced anxiety. Where was what? She did not have any secrets—or, rather, any secrets that these searchers did not already know.

She lifted her discarded multicard and called up the detective's number. She tapped it over to the phone. "Detective Roberts? It's Carolyn Gray. They've been in my office."

"Sit tight, Dr Gray. I'll send Forensics over."

*   *   *

Forensics shooed her from her office. She wandered back down the two flights of stairs to the lab level and coffee room. It was not until she reached inside the cabinet, searching for the hot chocolate that Frank and Aya had been drinking, that she realised she still clutched the note in her left hand. She should bring that back. She turned slowly, trying to think how she would explain the name. Everyone would be talking about this. The police might not make the connection to Baby C, but surely several colleagues would.

Detective Roberts entered the coffee room. "We've got some footage I'd like you to look at."

Carolyn sat beside her in distracted relief. "Sure." The tiny, no-nonsense woman was like a stabilising force.

Roberts set her tablet upright on the low table and tapped the screen. A black and white image showed the front of the building from a steep angle. "Whoever did this disabled video in and around the building, but they missed this ancient camera a block away." Three men approached the doors. Two wore overcoats and had dark hair, cropped short. The third wore jeans and a puffy jacket, and had scruffy, greying hair with thick sideburns. One of the younger men stood by the card swipe for a few seconds. Then the doors opened, and they chivvied the third man inside. "Do you recognise any of them?"

Carolyn frowned. Something about the older man seemed familiar. But perhaps it was just his last-century hair and its eccentric, academic appearance. "The old guy … Don't you have face recognition?"

"Nothing has come up yet. The resolution on this is horrible." Roberts rewound the clip to the best shot of the old man—a bit more than a profile.

Carolyn squinted, trying to place him. "Sorry."

"Carolyn?" Frank leaned around the door frame. He grinned as if this morning had not happened. "I did some modelling. I know where the petites came from."

Roberts handed Carolyn a business-card sized docfilm with a capture of the face from the video. "Let me know if you place him." Carolyn stood and stepped toward Frank, then looked back at Roberts. Roberts waved her on. "We're done for now."

*   *   *

Dark had fallen by the time her discussion with Frank finished. His partial construct was answering a completely different question than his PhD research. They might even get a paper out of it. She went back to her office, reflecting she had accomplished only one of her planned activities for the day. Although trying to understand who had turned her lab upside down and a left mysterious note in her office had not been on her list.

Mysterious note. She jammed her hand in her pocket and fingered the rough paper of the note. Forensics was long gone, and it was well past working hours. Plus, she was starving. She would get it to someone tomorrow. That the decision put off explaining her old name just a little longer was not the reason, she rationalised.

Chapter Two

Carolyn woke from a dream of chasing a side-burned man who had stolen Tonya's paper and turned it into DNA. The telomerases were not working, and each time he photocopied it, it got shorter and shorter until there was nothing left. She sat up in the morning dim, her heart thumping.

She stumbled out of bed and through her morning routine. She banged on her daughter's bedroom door. "Ellen! You up?"

Sleepy moans filtered through the door. She cracked the door open to see the tousle-headed thirteen-year-old push herself to sitting. "You look like I feel," she told her daughter.

Ellen tossed a plush rainbow heart in her direction. "Mum!"

"I've got lecture again this morning." Carolyn's brain moved slowly; there was something else about today and Ellen. *Right!* "Remember you're going to your Dad's after school. Do you need help with your lunch?"

"No, I've got it." Ellen laboriously dragged herself to her feet. She shouldered her way past Carolyn to the bathroom.

Carolyn was nearly done with her coffee by the time Ellen plopped down at the breakfast table with a bowl of porridge. Carolyn took a last lukewarm sip and stood. "Have a good time with your Dad. See you Saturday." She leaned down and hugged Ellen. "Love you."

Ellen returned the hug. She tilted her head. "You okay, Mum?"

"Fine." Carolyn had not said anything about the lab vandalisation, but perhaps Ellen had picked up on her foggy preoccupation. "Just work stuff." Ellen gave her an unusual extra hug.

*   *   *

Carolyn spent her tube ride studying the docfilm of the old man, instead of reviewing her lecture. She wondered if there would be answers today.

In the lecture theatre, she rubbed her eyes and tried entering her University password for the third time. Finally it worked. She flashed back to her dream as she stumbled through an explanation of telomerase. She hoped the students at least got the concept that it did not stop the shortening, but instead extended the buffer.

As she retraced yesterday morning's walk through the windy campus, her stomach flipped with nerves. Perhaps today she would get further into her list of activities. The building was on card key, but no one guarded the door.

She poked her head into the lab. Aya, Mohan, and Tonya bent over their lab benches. Mohan gave her a thumbs up. She grinned back. His assays must be going well.

Tonya, facing Mohan, turned at his gesture. "Have you read my paper?"

"No—it's on my list!" said Carolyn. Tonya turned back to her tubes. It appeared no one else needed her. Carolyn continued on to her office. That was how yesterday should have gone.

Her phone was blinking. She stared at it for three full blinks before remembering that this was the indication for voicemail. First cursive, now voicemail. It felt like antique communications week. Her hand sweated as she picked up the handset.

Beep.

"Dr Gray, this is Hugh Nguyen from Sandslin Corporation. We understand you had a disruption in your research. Please contact us at your earliest convenience." His tone suggested 'earliest convenience' meant 'now'.

Three more nearly identical messages followed from the remainder of her funders. News travelled fast. She moved to put the handset down.

Beep.

She brought it back to her ear. A breathy male voice said, "Carolyn, I'm sorry, they said they left it with you. Just hand it over, or leave it out, or, or something. These people … just give them what they want. I …" The message ended.

Queasiness settled into her stomach. She stared at the handset as if it could reveal something. She had assumed whatever was going on was something to do with research, or her funders, or something less personal. But the note, and now this. Who was this man? *Who* left *what* with her?

The phone rang. She toggled the handset. "Hello?"

"Dr Gray? It's Hugh Nguyen, I left a message."

"I just got in." Her heart was still speeding from that last message; it was difficult to think straight. "The lab was messed up, but we're back at work today. There shouldn't be any effect on our research."

"That's good to hear."

"I can give you the contact for the detective—"

"Just keep us informed. I understand it was a student prank?"

She wondered how he knew that much already. "Actually, maybe not."

"So it was really a search for something? Do you know what they were after?"

"No. Do you?" It was a facetious reply, she knew as she said it. That last message still had her off balance.

"Not the Schwarz Final Findings?" he said with a laugh.

"Why would you say that?" Her tone was too sharp. Sandslin did not know about her former last name. There was no reason for them to jump to the legend of her parents' mysterious last research.

He laughed again. "Oh, just that they spent some time at LSU. I can't imagine why else someone would ransack a genetics lab."

Her parents had worked *here*? She had not known that. "I study mitochondria, not genetics." He should know, if he was calling about their funded project.

"It's all the same to me. Anyway, keep us informed." He hung up.

She stared at the handset, again wanting answers, still reeling from the idea she had ended up at the same University her parents once had … what? Worked? Studied? She had thought they had gone straight from postgrad to corporate work, but she had purposefully avoided following their careers in detail. Her stomach twisted in a knot. It could not be the Schwarz Final Findings the vandals wanted, could it? That was a myth. Decades ago people had believed her parents had made a major breakthrough right before they died. But that was back before Baby C—Carolyn—was revealed as a hoax.

Yet the searchers knew her name. She rubbed her hands. The police did not know that. She should call the detective about the note and the message. She did not want to. A simple solution had not presented itself in the intervening day—as she recognised she must have been hoping: if she could 'forget' about the note for just a while, perhaps the whole mystery would be over before she had to explain her other name.

If she waited much longer it would be embarrassing, like she had something to hide. She steadied her breath. The police were unlikely to recognise the import of the last name Schwarz, she rationalised. Maybe her colleagues would never hear. She called up Roberts' number on her multicard. She stared at it for one last second, breathed deep, then tapped the number over to the phone.

"Detective Greg Walsh," a male voice answered.

"I'm looking for Detective Susan Roberts," Carolyn said.

"She doesn't work here anymore."

"Um, who's handling my case? I'm Carolyn Gray."

"I'm sorry, I'm not familiar with your case. Let me patch you through to the front desk."

Several transfers later, and still no one had heard of Carolyn's case. She left her contact information and hung up slowly. That was odd. Roberts had not given any hint that she would be leaving. Something must have happened. An accident? *She doesn't work here.* Fired?

Carolyn chewed her lip, wishing for the comforting voice of the small, no-nonsense woman. To her, Carolyn's lab disaster was just another day at the office. Not this extraordinary event littered with unsettling details. She found herself hoping Roberts had been fired, rather than some horrific accident. Could she be dead? They would have said ... maybe not.

She pushed away her increasingly paranoid thoughts, recognising a tendency to catastrophize. She had managed to keep from doing so for the incident in the lab—although partly because she could not even guess a worst-case scenario to imagine—she need not complicate things with being anxious over events on the edges. She and the detective had crossed paths for less than a day; they were essentially strangers. If not for the coincidence of her case yesterday, Roberts' tragedy-or-not would have played out with Carolyn completely unaware. Someone else would take over her case, and she would have another no-nonsense detective to provide balancing perspective.

She opened her email to contact the rest of her funders. She could not face more phone conversations right now. Then there was still a pile of work to do, double the load due to the disruption yesterday.

*   *   *

The police front desk still could not find her case. Carolyn put the handset down. She had finished most of her tasks regarding the disturbance yesterday, reassuring her funders. But there was still the matter of the note and now also the phone message. She had finally taken the plunge to do something about her embarrassing reticence, and it was not working. How could her entire case have vanished? There had been masses of people here yesterday. Most were forensics. She turned to her computer and found the direct number to the campus forensics department.

It rang seven times before a woman picked up. "Hello?"

"This is Dr Gray from Building Seven. I'd like to speak to someone who was at my lab or office yesterday."

"That won't be possible."

"Why not?"

Scraping and feedback came over the line, as if someone were covering the receiver. "Hey, I'm using this!" said the woman who had answered, faintly. "I'm sorry," she said more clearly. "Things are a bit chaotic here, what with all the new people. Let me take your name and number, and someone will get back to you."

"I have new information." Relief flooded her: first time she had said that.

"I can't do anything with it now. What about your detective?"

"They've lost track of me too. Please, I just want to speak to someone involved in my case." She made a fist. Why was this so hard?

More scraping sounded, and unintelligible speech. "I truly am sorry. Let me get your info?"

Carolyn gave her name and number. Her mouth was dry. *These people … just give them what they want.* Her mysterious vandals could not have something to do with whatever was going on at the police, could they? She dismissed the thought.

But, as the day went on, it nagged at the back of her mind. Nguyen's joke about the Schwarz Final Findings preyed on her as well. Someone involved in the break-in knew her original last name.

By the time evening fell, she was dropping styluses from nervous fingers. She should get some sleep. It was a child-free night; she could eat dinner while watching some soppy movie without fear of teenage ribbing. Tomorrow the police would surely have recovered her material. She pulled on her coat.

Outside, the campus was empty, but scattered bright windows of postdocs or postgrads working late revealed bustling energy inside. The wind was nearly as strong as the morning before, chilling her cheeks and trying to sneak down her neck. She pulled her coat tighter about her. She would appreciate a nice, warm evening in front of the 3D.

Crack.

She froze, heart thumping. An elly-book skidded along the side of a building, propelled by the stiff wind. She laughed in relief and ran to catch it. Probably an e-lab book that had somehow gotten misplaced. Someone would sorely miss this if it got eaten overnight by the campus autosweepers; she could return it tomorrow.

Her relief faded by the time she reached the edge of the University buildings. Sounds oddly muted by the high wind triggered her overactive imagination. She wished she had spoken to Roberts today. The detective's stabilising force would have calmed her racing mind.

Footsteps sounded behind her. She turned. No one was there. The winds could have carried sound from anywhere. She stretched her stride, eager to reach the tube and the bustle of late commuters, to normalise the evening.

A sharp tug on her elbow set her stumbling sideways into an alley. Hands grabbed her arms and threw her against a wall.

"What the hell are you mixed up in?" said a small, shadowed figure.

Chapter Three

Carolyn looked down into the snarling face of the no-nonsense woman she had just wished for: Detective Roberts.

"Noth … nothing," she stammered.

"Tell me the truth!" Roberts jammed her harder against the rough brick.

Fear spiked through Carolyn. "I don't know!" She shifted her shoulders, trying to loosen Roberts' pinching grip. The detective's eyes looked wild in the dim light. Her hair hung in scraggly patches, half pulled from a ponytail set asymmetrically and sagging. The left side of her face was developing a bruise. "What happened to you?"

"Like you don't know!"

Blood rushed in Carolyn's ears. Stress and anxiety burst into fury. "Where have you been? What's happened to my case? My funders are acting bizarre—they *phoned* me. There was a note and a message …" She was being incoherent.

Roberts took a step back. "You really don't know?"

Carolyn shook her head, fury draining.

"Whoever ransacked your lab tried to have me erased. I got away."

Carolyn's stomach dropped. "Erased?"

"Like I never existed. I've got no job, no flat. I barely have my life. And it's all your fault."

"Someone tried to *kill* you? It's not my fault!" But if it was related to the lab … her mind took a jump: *Ellen*. Could whatever had happened in the lab somehow impact her daughter?

"Like hell it's not!" Roberts was wild-eyed again. She propelled herself at Carolyn, hands out like claws. Carolyn stumbled backwards, but Roberts dropped her arms halfway to her. "You're so damned innocent." She tilted against the opposite wall and laughed. Her legs folded, and she bent her face into her hands, still laughing. "So damned innocent."

Carolyn needed to phone her ex and ensure Ellen was safe. She sidled sideways experimentally. Roberts kept laughing. Carolyn stepped towards the street.

Roberts sprung to her feet. "Where are you going?"

"Home."

"You can't go home." Roberts reached for her arm again, but Carolyn jumped away. "They know where you live."

"How could you know that?" Was Roberts delusional?

"They knew where *I* lived."

"Who are 'they'?"

Roberts lunged at her. "You tell me!"

Carolyn ran. She fumbled her phone from her pocket. Nicholas's number went straight to voicemail. "Call me!" she shouted into the recording, panic rising. She fired off texts to Nicholas's and Ellen's phones. Ellen would be safe—she would have gone to school and to her father's afterwards; no reason to cross paths with murderous vandals. If they even existed. *No need to panic*, she repeated to herself like a mantra.

*   *   *

She opened her flat door slowly, hyper-aware of her surroundings. *They know where you live*. What if it was more than paranoid ravings?

But the flat looked fine: leather couch with its rumpled tartan throw; kitchen with Ellen's breakfast plates still on the table. Automatic, mild annoyance flitted through Carolyn's mind, but could not take hold. The door to Ellen's room was open, and the light on.

A soft noise came from farther in the flat. Carolyn swung her knapsack around to her front like a shield. More noises. They came from Ellen's room. She wanted to back out the door, but if someone was here, and in *Ellen's* space … she had to know.

She crept forward. She paused, heart thumping, then peered into Ellen's room.

Ellen lay on her stomach on the bed, propped up on her elbows with knees bent and feet crossed in the air, looking at an elly-book between her arms. Carolyn breathed out in relief.

The screen of Ellen's elly-book showed a swirling, cloud-like image. Ellen tapped, and four numbers filled a blank box in the screen's centre. The screen spun, the numbers stretching as if they were flowing down a drain, the initial two two's extending into thin fishing hooks. A new, split screen appeared: plain black on the bottom and text bubbles on the top.

Ellen sat up abruptly and pulled a pillow over her elly-book. "Mum! Why'd you sneak in like that?"

"What are you doing?" Carolyn had never seen anything like that on an elly-book. It was meant to be just for school notes.

"Nothing?" The lilt to Ellen's tone indicated she knew the denial would not work.

Carolyn sat next to her on the bed. "It's okay. I'm not mad." She was just glad Ellen was not an intruder. But what *had* she been doing? "Why aren't you at your Dad's?"

"He had some kind of dinner thing. He said to come later."

"How much later?" She did not like the idea of Ellen wandering London this late at night.

"Tomorrow?"

Carolyn pushed away familiar frustration. She was getting distracted by Nicholas's change of plans. "What's with your elly-book?"

"Um." Ellen ducked her head. "I was chatting with Jean. Sorry, Mum, but it's safe—it makes a tunnel and is completely untraceable! No one can listen in, at all. You can surf the web, but it's only text. All the kids are doing it …" Her voice trailed into a squeak.

They had talked about internet safety; Carolyn had thought Ellen was on board. She was meant to only use the highly regulated app on her phone. "You don't know what's out there—"

"No, Mum, I know. I only chat with my friends. I know it's them. You need two elly-books, you see. You've got to know their code. Then both of you can surf the web. But I don't use it for that, we just chat. Some kids go to these message rooms, but I knew that would be dangerous. I just wanted …" She hunched her shoulders.

Some privacy, Carolyn mentally filled in for her. It had been the same when she was young: new apps that the adults did not understand, something where your parents—or uncle—could not scroll through your text history. But the world was so much more dangerous now.

Carolyn put an arm around Ellen. "I'm just worried about your safety. You don't know who might be listening in."

Ellen sat up straight, animated. "No, that's the thing, Mum! Really, no one can eavesdrop. There's them that use it for anticorp stuff. But I don't," she hastened to add. "That crowd's scary! Elly-tunnels are completely secure. Aunt Bae said not even the corpsies can watch you."

Corpsies: corporation spies, Carolyn mentally translated. In a world where anything you wrote on docfilm was stored in a distributed computer system owned by some company, and every text and phone call was recorded and stored on some other company's computer system, it was paper or elly-books for the privacy-conscious. Elly-books made good e-lab books because they were self-contained, just like paper lab books before them, not hackable by a competitor—not that any of Carolyn's research was in such danger. She had her lab members write on docfilm, too, for both the immutable record and the data backup. But for some privacy mattered more. An elly-book was overkill for Ellen's school notes, but Nicholas had gotten it for her. They had only rudimentary internet connectivity. Even the games Ellen liked needed to be plugged in physically. No data transfer was meant to be possible. Yet someone had clearly figured out how to do it, if in a very primitive manner.

At least if what Ellen said was true, no one should have been able to listen in to a conversation from Carolyn's flat tonight. Carolyn hugged Ellen.

"Are you angry?" asked Ellen.

"I'm disturbed you didn't tell me," Carolyn said. "Even if it's safe from eavesdropping as you say, it's dangerous to hide things. Hiding stuff gives leverage to anyone who finds out—remember we talked about this?"

Ellen nodded solemnly. "Right, Mum. I'm really sorry."

"It's okay." But Carolyn's mind was elsewhere—hiding her parents' identity had given leverage to the vandals. She had not shared the note with Roberts when she had the chance, because she had wanted to keep her secret. Her true last name had seemed a simple thing to hide, to her teenage self, when she never intended to become a scientist. She gave Ellen another squeeze. "I trust you."

Ellen leaned back and crossed her arms. "So why'd you go through my stuff?" Her tone turned from contrite to angry.

"What?" Carolyn, blinked, confused.

"My stuff. You went through my desk. When I got home all my papers were ruffled. You said you'd *never* do that, if I told you stuff."

Carolyn ignored the blatant fact that Ellen had *not* told her stuff. "El, I just got home. I left before you. I didn't go through anything."

They stared at each other. Carolyn's breath caught. They'd been here! She jumped up and ran the rest of the way down the corridor, throwing open doors. Her study, bedroom, and the bathroom were empty. Only as she reached the end of the corridor did she realise how reckless that had been—what if she had encountered someone, with Ellen in the flat?

She went to the one place in her home she kept things from work: her study. Her desk drawers were partially open, papers sticking out. She had not left it like that. She backed to a wall and wiped her palms on her legs, as if she could wipe off the feeling of someone else.

"Mum, what's going on?" Ellen stood in the doorway, tense.

Carolyn controlled her panicky breath. What to tell Ellen? *How* to tell Ellen? Carolyn did not even understand!

She could approach this like unexpected lab results. What did she know? Some people—the Vandals—were looking for something. She had been left a note and phone message: to apply Occam's Razor, she would assume they came from the same person, someone trying to communicate. She would call him the Informant. Someone had told *him* that the object of the Vandals' desire had been given to her. The Informant also knew her original surname.

The Informant must have known her parents. She fell into her desk chair. It must have been they who gave her something. Maybe even the Schwarz Final Findings. She shoved her fist into her mouth to mute a nervous chuckle.

"Mum?"

Carolyn sprung to her feet. Legendary missing research aside, she had a place to start with her parents. First stop would be Uncle Keith in Scotland. "There was a break-in at the lab." She got a knapsack and began filling it. "I think they came here, too. The police ..." She trailed off. She had encountered the police on the way home, sort of. "I'm not sure who to trust. I'm not comfortable staying here this weekend. We'll go to Grandpa Keith's."

"I go to Dad's tomorrow."

"You go to Dad's tonight." The Vandals had left, but would they stay away? She wanted Ellen somewhere else *now*. "Get your school stuff and pack a weekend bag. You can take the train up on Saturday ..." Ellen had done that before: they would often spend a bank holiday weekend in Scotland, and if it was a week Nicholas had Ellen, she would join Carolyn partway through. But this was not just a holiday weekend. She was nervous about Ellen on the train alone. "No, I'll come get you. Or Grandpa, or Aunt Bae. Or Dad will take you up. Don't travel on your own, okay?"

Carolyn felt more in control. She would go up after work on Friday, look through her parents' stuff, and figure out what was going on. By the time Ellen got there, they could have a pleasant weekend visit.

"Mum, you're scaring me." Ellen had gotten down a bag, but stood with it clutched to her chest.

Carolyn hugged her. "It's probably nothing. But let's just be safe."

"Shouldn't you call the police?"

"I'll call them after we get you to your Dad's." She would not, though. She did not trust the police anymore. Her mind supplied a vision of Roberts, face bruised and hair dishevelled: perhaps fixing this would not be so simple as she hoped. She pushed the image away. If these people wanted something of her parents', she could just give it to them, like the phone message had said.

Chapter Four

Squeaks and clatters filled the lecture theatre as students pushed down their desks and gathered their belongings. Two students in the back of the room caught Carolyn's eye. They were not bending over to grab coats and shoving things in bags like the rest. They stood, hands empty. The pair were oddly uniform: black trousers and tight grey sweaters.

She was suddenly certain that these were the two men who had entered her lab. She sped through the logout process on the terminal. One of the men climbed over the seats in front of him, no longer waiting for the students to clear.

A brown-haired young woman in a flowered skirt, hugging an old tablet, approached the lectern. "Dr Gray—"

"Send an email." Carolyn grabbed her own coat and the knapsack at her feet. She slipped into the student exodus, heart thumping. She looked over her shoulder as she moved through the door. Both men were now clambering down the seats.

She broke into a jog as soon as the corridor allowed, struggling with her coat. It seemed to have been made for a stubby-armed octopus. She hugged it to her chest and kept going.

Maybe they just wanted to chat. She could have waited and explained she did not have whatever it was they wanted. Her mind flashed to the memory of Roberts' wild eyes and dishevelled hair. She sped to a run.

She burst through the doors to the frigid outside. She attempted her coat again and found the sleeves. She could not go to her lab. She could not go home. She could not go to Nicholas's and make Ellen a target. She turned sharply, a tube station her vague goal. She could get to the rails and take a high speed to Scotland; her plan for today if not originally right now. If her uncle's attic held whatever item her parents had left her, she could simply deliver it to these people and be done with it.

*   *   *

She fidgeted in the ticket queue. She checked the doors behind her again, but no pair of grey-sweatered men had entered. Finally she crossed to the woman behind the window. She slid her card under the low slot. "I'd like—"

"Two tickets to Paris," said a voice at her left elbow.

Carolyn stared at Roberts, suddenly beside her. She opened her mouth.

"Pay," said Roberts under her breath.

Carolyn turned back to the finger pad the woman passed back under the slit along with her card. She held her index finger on the pad until it beeped.

The ticket woman caught several slips of docfilm ejected by a printer to her left. "The next train leaves track E-twelve in forty minutes. E-level has its own lounge." She pushed over the docfilm stack.

Carolyn took the tickets and let Roberts lead her aside. "Where did you come from?"

"You're ridiculously easy to tail," said Roberts. "I wasn't the only one doing it, either."

"But I didn't see—"

"Do you think those two are all they have?"

Oh. Carolyn stopped, and Roberts tugged her on. "Why Paris?"

Roberts stopped to look at the departure boards. "Someone *else*. Where were you headed?"

Carolyn stared down at Roberts. Her ponytail had been tightened to its previous smartness, and a thick foundation mostly obscured the bruise. *Some*one *else?* "Stirling. My uncle lives there. I hoped I'd find—"

"Your uncle? Surely they know …" Roberts hunched and pressed her lips into a thin line, still scanning the boards. She straightened. "Well, at least it will confuse them." She held out a hand. "Ticket?"

Carolyn passed her one. Yesterday, she had wished for Roberts' steadying presence, then run from her apparent dramatic change. But here was the no-nonsense competent detective again. "I can't go to Paris." Ellen was expecting her in Stirling this weekend.

"We're not." Roberts led her through the ticket barriers and into a ladies' room. She pushed Carolyn into one stall and entered the next one. Carolyn closed the door to her stall.

A hand appeared underneath the wall separating the stalls. "Ticket and all receipts."

Carolyn fumbled them out and dropped them into Roberts' hand. The hand retreated.

"What did you want to find in Stirling?" The wall creaked as if Roberts leaned against it.

"Hopefully whatever they want."

"What's that?"

"I don't know!" Carolyn could not help the note of petulance that crept into her voice. She bit her lip, wondering at Roberts' reaction on the other side of the wall. "Something my parents had. That's all I know, I swear." Roberts' continued silence unnerved her. "Why are we in the toilets?"

"No surveillance in the stalls. But we can't tarry too long." The hand reappeared holding a small stack of docfilm. "There's a Stirling train at eleven thirty-two run by Grampian Rail. We'll take that."

Carolyn retrieved the stack. "There's an earlier one by Coastal." She preferred to avoid run-down Grampian if possible.

"We want Grampian."

"Why?" Her ticket had the pink cast of disconnected docfilm, yet the text inexplicably remained. She blinked. No, it was *different*. She now held a ticket to Stirling. The pink faded back to grey, but the changed text remained. She forgot about all questions of train times. "You *hacked* docfilm? I didn't know that was possible." She trusted the integrity of docfilm for her research. If they could be forged … people would be even more suspicious of her work if her parentage came out.

"You learn some useful things in the force." The toilet in Roberts' stall flushed. "Use the facilities if you need."

*   *   *

Carolyn followed Roberts on what seemed an aimless wander of the station. She was bursting with questions, but kept silent, remembering what Roberts had said about surveillance. They went to and from E-level, and Roberts got her to take out the max allowed from her account in cash.

"Off the grid," Roberts said softly as Carolyn fingered the unfamiliar paper, slick like docfilm but not as thick. She had not held notes since she was a teenager. She handed half of it to Roberts; it did not seem like real money. As they walked, Roberts took networked devices off Carolyn one at a time, surreptitiously ran something over them, and handed them back. "Don't turn *anything* on." Carolyn nodded.

Roberts finally collapsed into a chair in E-lounge. "I think our friends have decided to go on ahead and meet us in Paris."

Friends? "Someone's been following us?" She was catching up, if slowly.

Roberts gave her a withering look. "Quiet for now. Get some rest if you can. We'll talk on the train."

"Will it be safe there?" She pushed away a growing paranoia. Roberts knew what she was doing. She would trust her until she was in a position to demand answers.

"Shh!" Roberts closed her eyes. Carolyn squinted at her. Was she actually sleeping? Carolyn closed her eyes, but snapped them open immediately. Roberts might be ready to relax, but Carolyn could not. She hugged her knapsack and watched the large analogue clock above her.

Chapter Five

Roberts pushed herself back up in the small train seat. She was perched on the fold-down extra seat in the small end of the carriage beyond the toilets; the seat's attachment to the wall was decidedly questionable. Carolyn sat across a half-table from her, in the one full, if lumpy and threadbare, seat in the area. A nearly overpowering scent of lemon-mixed-with-bleach reassured Carolyn that the train car was cleaner than it looked.

"What's this about your parents?" Roberts asked.

Carolyn looked around. She knew very little about surveillance, but she figured it had to be mounted on something, like walls. The train car's dingy blue walls were much closer than those in the station. "Are you sure we can talk?"

Roberts patted a pocket. "Dead zone. At least half of Grampian's are always out."

"That's why we waited?" It had not occurred to Carolyn that the poor reputation of the tiny rail company would have benefits for them.

"Yes," Roberts said. "Plus, they're an independent, so any recordings they do make won't be shared with London PO as a matter of course."

"Huh?"

"You must know that all the police in London are owned by the same corporation?"

"Um …" Carolyn had a vague concept that the campus police was somehow connected to the various branches of the city police.

Roberts leaned back, then quickly steadied herself in the seat. "London police—in fact most of the South—are owned by one of the Big Twenty. We get all surveillance from their subsids, like the transport companies, piped straight to the analysis guys." She wiggled her fingers, then made a fist. "The speed at which I was ousted, and the folks at Forensics, indicates that whoever is after you is owned by them too."

"The criminals and the police work for the same people?" Carolyn felt her worldview slipping. Surely a corporation—or at least one that was the *police*—would not commit a crime. Her mind conveniently supplied memories of news stories of scandals and illegal activities … but that was all corporate stuff, involving vast sums of money and massive operations that had little relation to an individual's life. Not breaking into a random scientist's lab. Besides, she had not seen stories like that for ages.

Was that because the criminals now owned the *police*? Her stomach twisted. "Was the mess at forensics,"—*all the new people*—"due to me too? But it doesn't make sense. If the Vandals own the police, why would there even be an investigation? Or an incident in the first place?"

"It could be the search couldn't be inconspicuous, so they went the other direction." Roberts tapped the small half-table with a stylus. "We were directed heavily to the student prank theory. It was only your phone call about your office and that ancient camera that suggested otherwise."

Carolyn sat up straight, recalling something else the detective did not know. She pulled the note in its envelope from three days ago from her pocket. "I forgot to mention … This note was in my office that first day. I meant to bring it back to Forensics, but I never got around to it."

Roberts read the note and handed it back. "Schwarz?"

"My parents' name. They died when I was young, and I took my uncle's name."

"Is that why you think this has something to do with your parents?"

Carolyn nodded. "I also got a phone message. The fellow specifically said my parents told him they left 'it'—whatever that is—with me."

Roberts flipped the stylus over her fingers. "You never got this note to Forensics? Did you tell anyone?"

"No."

"Good. What about the message—did you delete it?"

Carolyn reached for her phone. She had remote access to her University voicebox. "I could—"

"No!" Roberts grabbed her arm.

Right. Keep the electronics off. "Did you disable some kind of tracking?"

"Yes." Roberts patted her pocket, perhaps touching the item that had done it. "They'll know I'm with you, since this is police tech. But they saw us together anyway." She rested her chin on steepled hands. "We can hope they don't check your phone messages; it's such an old-fashioned way to communicate." She said the last in a tone suggesting everything about Carolyn was old-fashioned. Carolyn opened her mouth to defend that it was other people who had left the messages, but Roberts spoke before she could. "Do you have any idea what they could be looking for?"

Carolyn laced her fingers together. "There is a legend about my parents. They were supposed to have found some big thing before they died."

"A valuable object?" Roberts crinkled her brow.

"No, science. It's called the Schwarz Final Findings." It sounded ridiculous to say out loud. She cringed in embarrassment.

But Roberts just kept flipping her stylus. "What area?"

"I don't know. I would guess synthetic genomes. That was what they were famous for."

"Carolyn Schwarz …" Roberts said softly. She put down her stylus. "Wait, you're not *Baby C?*"

Carolyn's face heated. "You've heard of that?"

"Of course! It was all over the news when I was a teenager. I remember feeling bad for her. She must have been near my age; it seemed like it would have been a nightmare."

"It was."

Roberts grimaced. "Sorry." She flipped her stylus over her fingers. Twice, she looked about to say something before she finally spoke. "This legend: what do you know?"

Carolyn shifted uncomfortably. "Before the hoax was discovered, it was part of their mystique. Some said it had something to do with why they died. I must admit I didn't explore it much—my uncle kept most of it from me. By the time I wasn't a kid anymore, we all knew they'd lied about me. So I assumed the Final Findings were a lie too. And that everybody else did."

"Topic? Any hint we can go on?"

Carolyn shook her head.

"Who did they work for?"

"Vivcor."

Roberts straightened. She gripped the table, rebalancing. "Vivcor had a major reorganisation recently. Nearly went under. Perhaps something got unearthed in that process?"

Vivcor had helped Carolyn hide her embarrassment; she did not want to think bad about them. "Are they part of the police corp?"

"Not to my knowledge." Roberts tapped her stylus, staccato. "But it was like vultures at a Vivcor carcass. Who knows what subsid got access to what."

"Well, that's a working hypothesis." Carolyn felt better if she could build a story behind what was happening. It was not Vivcor's fault if someone had found something they did not understand. "Something was discovered in Vivcor's reorganisation that made someone think my parents really did have some final findings. No idea why they think I have it—"

"Because that fellow told them …"

"Right. So my parents told him they gave me something." She crossed her arms again. "One of my funders said they used to work at LSU, too. Maybe it's not me but the place?"

"Your funders said … Don't you know?"

Embarrassment about her lack of knowledge built. "I don't know much about them. They died when I was very young. Even before the hoax was discovered, I didn't like the attention of being *different*." She felt like she was making excuses. But it had not mattered until now. "I hope my uncle has whatever it is."

"Me too." Roberts was silent for a while.

Carolyn had been brimming with questions before, but she could not think of one now—other than, *What is going on? This isn't real, is it?* Her mind flitted to Ellen, who would be in school. Carolyn needed to sort this out and make sure none of these surreal events impacted upon her daughter.

Roberts cleared her throat. "I hope you don't mind me asking … I never really understood how they knew Baby C was a hoax?"

"Three X chromosomes." Carolyn did not want to talk about that.

"Yeah, I remember, but why does that mean you couldn't've been synthesised?"

"Most people have two of each chromosome, except the sex chromosomes, where women have two X's, but men have an X and a Y. You know people are made from egg and sperm with one set of chromosomes each?" Despite herself, Carolyn began to relax as she turned on teacher-mode. "Sometimes there is an error in the making of an egg or sperm. It's known as *meiotic nondisjunction*. Basically, when the—let's say it's an egg—is made, there's a stage where all the pairs of chromosomes line up, and one each goes into a new cell." She pulled out the docfilm notebook still in her bag from yesterday's lecture, intending to sketch.

Roberts put a palm on it, holding it closed. "Don't contact the cloud."

Carolyn leaned back. Oh. Docfilm was not in constant contact like other devices, but writing on it activated it. She held up her hands instead, palm to palm to simulate the pairs of chromosomes. "But sometimes something goes wrong, and instead of one of each pair going into a new cell,"—she separated her hands—"two go into one, and none into the other." She brought her palms together again and shifted them across her body still touching. "If that happens, the egg would now have two X-chromosomes. If it meets with a normal X sperm, the resulting child has three."

"Oh," Roberts said. "So it's not like someone couldn't synthesise three of them. Just that it must have come from a real egg. I always assumed it was just too much to synthesise."

Carolyn laughed, with bitterness. "Turns out the whole thing was too much to synthesise back then. But, no. The X chromosome is pretty sizeable, but it wouldn't make a difference once you could do a whole genome."

Roberts shifted, looking uncomfortable. "Um. Isn't extra chromosomes pretty bad?"

Carolyn leaned back. "The sex chromosomes are the only ones you can have an extra of without problems. Y is so tiny it doesn't do much but make you a man; they used to think that extra Y's made men more aggressive, but later studies put that to rest. For the X, because men only have one X, and women two, you really only need the genes of one. The other gets inactivated and forms what is called a *Barr body* in the cell, which is the X chromosome all lumped up. With an extra, you just get another Barr body. All your cells have one; all mine have two."

Roberts looked at her arm, almost as if she were trying to see inside the cells.

Carolyn said, "Funny story about how they discovered Barr bodies—"

A chime sounded. "*The next stop is Stirling. This train is for Glasgow, calling at …*"

Roberts stood. "Show time. Let's hope this is as simple as something left in your parents' old stuff."

Carolyn stood as well, butterflies filling her stomach. The questions she could not think of before now flooded in. *What do we do if they're following us? How would we know? I'm not putting my uncle in danger, am I?* She verbalised this last.

Roberts looked at her, almost sadly. "I'm sure they know who he is already. If it's there, we finish this. If not, we'll get out as fast as we can, and the trail should be on us."

That was not necessarily reassuring.

Chapter Six

The pale spear of the Wallace Monument loomed on its forested hill overlooking her uncle's house. The stone tower had a spiralling column running up the corner facing them; decoration-festooned arches rose from each corner and side, closing over the top like a giant eight-fingered claw. Carolyn recalled childhood hours of staring at it from her bedroom window, where it took places in her fantasies ranging from Rapunzel's tower to a spaceship ready to blast off. She ducked under the stone arch delineating the entrance to her uncle's garden. The tiny garden crawled with trellises sporting winter-browned vines; dormant potted plants nestled into corners and acted as end-posts for the trellises. As always when she visited, she was startled at how small everything seemed.

The front door opened before they reached it. In the doorway stood a small woman with salt-and-pepper hair pulled into a bun: Bae, her uncle's current partner. "Carolyn, what a pleasure!" Bae said with an open smile.

Carolyn shook off the sense she always had around Bae, that the openness somehow hid some nefarious plot, and stepped forward to hug her. "Good to see you. Is Uncle Keith home?"

"Having tea. Come in." She stepped back from the door and turned her face towards Roberts. "And this is?"

Carolyn took a breath, but Roberts spoke first. "Susan, Carolyn's friend." Roberts put out a hand. "Pleased to meet you."

Bae shook Roberts' hand, looking somewhat discomfited at losing the chance for another hug. She squinted at Roberts' face, then patted Roberts hand twice before releasing it. Bae retreated inside. "Keith! Your niece is here!" Carolyn and Roberts followed Bae to the sitting room, where Uncle Keith sat with a plate on his knees.

He moved the plate to a side table and stood. He took both Carolyn's hands. He exuded the rich, dill-heavy smell of cottage pie. "Carolyn, what's wrong?" Uncle Keith was always so perceptive.

Words burst from Carolyn. "Someone went through my lab. They were in my office, and in my flat! They even went through Ellen's things! There was a note, and a phone message …" She was leaving out the most important part. "Something my parents were meant to have—they're looking for it. I was hoping—"

"I've got all your folks' stuff." He patted one hand. "We can go through. No need to panic. We'll find it."

Carolyn's tension dissipated. Uncle Keith took everything in stride, from suddenly becoming a bachelor Dad of a toddler to calming a teenager who was convinced her life was falling apart. They would find whatever it was, and this week would be an amusing story to tell over coffee.

*   *   *

Roberts slid another box towards Carolyn. "More lab books."

Carolyn was looking through the science stuff, while Roberts viewed more varied personal items. The lab books were old-style, musty paper bound in blue or grey. These were the same as the last, dated some fifty years ago, probably from her parents' student days, given the detailed notes on things like how to run a centrifuge and pour a gel. Carolyn bit her lip, realising she did not even know when her parents had graduated University or gotten their PhDs. She had ignored too much about them.

Surely any 'final findings' could not be from PhD students. She flipped through the final book in the box, pausing briefly where, sandwiched between records of several long centrifuge steps, a double-page spread was turned over to a sketch of what must have been the lab. The ink drawing seemed like a window into a lost life. The lab was comfortingly full, with racks of hastily drawn test tubes jammed tight onto shelves, bottles with peeling labels on benches, and taped papers drawn with indecipherable squiggles representing whatever lab notes had been left between members. She found the familiarity oddly disturbing. This would be her parents before they had gone down the route of deception.

She scanned the final pages perfunctorily, then upended the box. A slinky-style rustle of clinks sounded. She lifted up two beaded chains containing laminated cards—ID cards. "Hey, these are more recent." The dates were a bit over thirty years ago, and the photos of a man and woman of approximately Carolyn's age, or a bit younger. Carolyn stared. Her parents had been younger than she was now when they died.

"That could be useful." Roberts reached across and took the cards. "Vivcor Paris. Don't think that's there any more." She flipped them over. "I wonder what this means."

Carolyn leaned forward to look. The backs of both cards were jammed with closely written letters and numbers, in marker above the lamination. "Some kind of code?"

Roberts pulled one over her head and offered the other to Carolyn. "Let's keep them."

Carolyn looked at the photo: her father. She had seen photos before, but of younger versions—when they revealed Baby C to the world. They had aged dramatically in the two to three years between that and these IDs. Or perhaps news releases were kinder than stark identification photos.

"Here's something." Roberts passed across a photo in a frame. It was Carolyn's parents again, considerably younger this time, at some sort of picnic under a pink-blossomed tree with three other people.

One man immediately drew Carolyn's eyes: he wore a puffy jacket and sported wild sideburns. He was younger than she had seen in the video, but his bearing was unmistakable. "It's the third man!"

"Do you know who he is?"

Carolyn shook her head. "Maybe Uncle Keith does. Any names?"

But Roberts was already pulling the photo from the frame to look at the back. Nothing written.

"Although it does not add much—we already knew he knew my parents," Carolyn said.

"No, that was speculation," Roberts said. "We didn't know Sideburns was your Informant, either. This makes that more likely."

Roberts was thinking more like a scientist than Carolyn was. Carolyn had jumped to conclusions with barely any evidence, something she would have never done in the lab. Was being a detective like science, but with everyday people? Perhaps she and Roberts had more in common than Carolyn had originally thought.

*   *   *

Carolyn passed the photo to Uncle Keith. "Do you know who the other people are?" Nothing of further import had appeared in the boxes; other than the ID cards, it was all from her parents' student days.

Uncle Keith squinted and turned the frame in the light, under the curious gases of Carolyn and Roberts. "Some other PhD students, I think."

"Did they stay in touch after they graduated?"

"I wouldn't know." He touched the photo lightly, stroking along his dead sister's hair. "The group at Vivcor was pretty tight. They didn't expand the socialisation circle much. I might have recognised one or two of them at the time, but not after these years."

"They studied in the States, University of Maryland, right?"

He made a hum of assent. "For their doctorates. They met as undergraduates—"

"At Cambridge," Carolyn finished. "I thought they went straight to Vivcor after that. When were they at LSU?"

He screwed up his forehead. "They weren't."

"Are you sure? An exchange, a placement: anything like that?"

He handed the frame back. "I would have known if my sister had moved to back to the UK. Why do you ask?"

"Why would your funder say they worked there when they didn't?" Roberts said at the same time. Carolyn frowned, unable to answer. Roberts asked, "Which funder?"

"Hugh Nguyen …" Carolyn said, remembering the name but not the company for the moment. She mentally reviewed her funders. All had left messages, but she had only spoken to the one. "Sandslin."

An indrawn breath sounded behind them. Bae stood in the doorway. "You take money from Sandslin?"

"They fund basic research," said Carolyn.

Bae pressed her lips together and backed away. Was she anticorporate? Did that explain Carolyn's impression of false cheer? "Does Bae—"

"Never mind," said Uncle Keith, not letting Carolyn finish her question. "Bae's views have nothing to do with …" He turned, as if he heard a sound Carolyn could not sense.

Bae's voice came from the corridor beyond: "You girls had better make yourselves scarce. Keith, the boxes!"

"Out the back." Uncle Keith's face looked worried, but with a purpose. Roberts jumped up immediately. She helped him guide Carolyn along. Uncle Keith said, "Tool shed, lower left. There's a hatch. Keep her safe." This last was directed to Roberts, who nodded seriously.

Carolyn's heart thumped as she followed Roberts through the overgrown back garden. "What's—"

"Shh!" hissed Roberts.

—going on? But Roberts wouldn't know either, for all that she acted as if she did. Was this what policing was like every day? It couldn't possibly be. And her uncle, and Bae: Carolyn felt rattled, like she had with Roberts in the alley. People she thought she knew were acting erratically. She glanced at Roberts, remembering that bizarre night and wondering why she had trusted her after that. Why she was trusting her now.

Roberts yanked open a door in the floor of the shed. She pulled Carolyn in after her. They lay down in a low but wide space, floored by cool dirt. The hatch shut with a shower of dust, and Carolyn squinted her eyes.

"Wha—" Carolyn stopped on her own this time, as the shed door opened. Heavy bootsteps clumped to a stop above them.

Chapter Seven

Lighter footsteps moved across the floor. Uncle Keith said, "I told you, there's just garden tools." The floor above Carolyn shuddered under heavy impacts; more dust showered down. "There's no need to make such a mess."

"Nothing," said a deep voice.

"I only have their personal things," said Uncle Keith. "Corporate stuff stayed at their work."

"Not even any IDs or prox cards? Things like that?"

Carolyn reflexively grabbed her father's ID card, still looped around her neck. Roberts touched her elbow, calming.

"No, I told you."

The heavy clumps left, followed by Uncle Keith's lighter step. Carolyn shifted, and Roberts squeezed her elbow. *Stay still*, Carolyn translated. Could they be looking for the IDs Carolyn and Roberts wore? Was that the same thing they were looking for in her lab, her office, her home? She had assumed it was something like a lab book or other kind of results.

It seemed an interminable, cold time before Uncle Keith returned and shifted whatever had fallen on the hatch. "There you go." He opened the hatch. Roberts sat up, and Carolyn copied her. She stretched her stiff arms and legs.

"Do they want these?" Carolyn lifted the card around her neck.

Uncle Keith's eyes widened, pale edges increasing in the dim. He touched the card and flipped it over. He squinted at the barely visible writing. "I didn't know you'd found this."

"Should we get it to them?" If all they wanted was these cards, maybe this whole thing could be over now.

He slumped in his squat, a defeated posture. "There's something you need to see." He stood and led the way from the shed.

Carolyn brushed off gritty dust and followed. "If I just get whatever they want to them—"

"It's not that simple," Roberts said.

"How do you know?" Carolyn's frustration focussed on Roberts. The detective had been in the hatch with her. She could not know any more than Carolyn.

They stopped in the conservatory, where Bae stood beside an ancient desktop computer situated in an alcove. The conservatory lights were off, leaving Bae's face to glow blue in the screen's light.

Bae stepped back as they arrived. "I'm sorry, Carolyn." For once, her voice did not give Carolyn that odd sense of falsehood.

The screen held a news article, headlined *LSU Geneticist Dies*. "Who?" asked Carolyn, her brain scanning through her colleagues. Bae's face was so sad; was it one of Carolyn's lab members? She stepped closer to see the body of article.

LSU Lecturer Dr Carolyn Gray née Schwarz was found dead in her fifth-story flat late this afternoon. Investigation is not yet complete, but the death appears to be due to natural causes. Gray held an impressive four research grants and was known internationally for her work on genetics of mitochondria, small DNA-laden power-centres of cells. Less well known, however, was Gray's original fame in the field of genetics as the Hoax Baby C of Drs Schwarz and Schwarz, the twenty-first century's Piltdown Man. "I would have never guessed she was Baby C," says Dr Kim Nixon, a colleague at LSU …

"I'm not a geneticist," Carolyn whispered, unable to formulate any other thought. The world felt as if it had fallen away. It was the Human Hoax all over again—except this time she was *dead*?

Her mind spun. "Ellen. What will Ellen think? We've got to let her know—" *That I'm not dead.* She finished the thought in her head, the absurdity of it unvocalisable. She should not have to be considering whether or not her daughter believed her dead. Anger built, like a smouldering coal bursting into flame. "Did they do this?" She rounded on Roberts, the one who always seemed to know more than she should. "Who in the world are *they*?"

Roberts ignored Carolyn's ire, staring instead at the screen. "It's your Vandals. It has to be." She lifted an arm to point at the article lower down.

> ... Gray had recently been the focus of a student prank which disrupted her research. "We actually like her," said Juliet Hu, a student involved in the prank. "We wouldn't have done it if we knew she was about to die." Hu along with three other students ...

"There was no prank ..." Carolyn trailed off, feeling adrift. Anger and unease intertwined. The Vandals had ransacked her lab, had chased her. The police—some of the police—were in cahoots with the criminals, generating a non-existent prank complete with student quotes. Generating her non-existent *death*. "How could anyone do this? *Why* would anyone do this?"

"I think—" said Bae.

Uncle Keith humphed. "It appears your parents may have really had something."

Unease turned to frustration at unanswered questions. "Had *what*?!" Carolyn shouted. Fury flashed about inside her, targetless. The others backed away.

She relaxed fists clenched tight and took a deep breath. Her anger settled on these *Vandals*. They wanted something her parents had? Screw that. She would find it before them.

She squinted at Bae. Resolve gave Carolyn confidence. She was going to sort this out. "I can't let Ellen believe I'm dead. But I can't let her be a target, either." The article had not even mentioned her daughter, which was oddly comforting. Whoever had orchestrated this did not think Ellen worth bothering with, for now, at least. She needed to keep it that way. "I told her someone would come down tomorrow and fetch her."

"We'll do that," said Uncle Keith. "But we need to get you somewhere safe."

"No. I'm not going to hide. I'm going to find my parents' Final Findings or whatever it is these Vandals want." She clenched her fists again. "Then I'm going to make sure it's either destroyed or brought into the open so that this secret can't hurt anyone else ever again." *Especially Ellen.*

"You don't understand what you're getting into—" started Roberts.

"I understand enough," said Carolyn. "I understand these people expect me to be scared and vanish. They've made a dire mistake. *They* didn't understand what they were getting into." She turned to Bae, who wore a growing smile. "I'm going to need your help."

Chapter Eight

The tall, bald man frowned at them. "I don't know, Bae. That one's a cop."

"Not anymore she isn't, Ron," said Bae. The glass-walled bus station seemed an odd place for a clandestine meeting, though the darkness allowing starlight to filter through the ceiling seemed to fit.

Carolyn hugged her knapsack to her chest. Her earlier conviction was fading into nervousness. Roberts had a bag now, too, filled with donations from Bae and Uncle Keith. It was almost vindicating to find that Bae's demeanour *had* hidden something darker. But mostly it was unsettling to realise she was throwing her lot in with anticorporate radicals. But she had little choice.

"I suppose being dead does make it more difficult," said Ron.

"They killed you off too?" asked Carolyn.

Roberts rolled her eyes, and Carolyn remembered Roberts' comment about innocence; she really was. "At least you died of natural causes. I either committed suicide or made a drunken mistake with sleeping pills."

"They must control the police to be able to find all these non-existent bodies." Carolyn frowned, infected by the distrust around her. Just how closely had Roberts been working with the people who ransacked her lab? But the investigation had seemed honest.

"Not as much as you'd think," Roberts said. "A person here or there. My death was snuck into the interstices of a complete change-over in forensics; I suppose they got the student prank evidence in that way too. Not sure how they managed you."

"That doesn't matter," said Ron. "Just how poorly attached are you to these people? Why wouldn't you give us up once this little situation is over?"

Roberts pulled herself up to all of her reasonably short height. "I became a cop because I wanted to help people—to find the truth. When I was a kid, cop didn't mean corporate. It never has to me. The opposite ..." She trailed off, almost as if she feared to say too much. But how could she, with this audience and their current actions? She appeared to come to the same conclusion, as she continued, "A cop's job is to uphold the rule of law, and the law's purpose is to protect *people*. Not corporations."

Ron crossed his arms and stared intently at them. Bae said, "Susan wasn't one of us, but she could have been."

He gave a nearly imperceptible nod. "We're nearing the end of our cover hour; we had best get you two settled." He turned to Bae. "They've got provisions for a day?" Bae nodded. He walked to the end of the bus terminal and crossed to the large garage where buses lined up. At the third bus from the left, he stuck a key in the side. The luggage hold slowly opened.

"Are you sure this is safe?" Carolyn asked, nervousness getting the better of her. Bae had told them how they would travel, but seeing the dark maw of the hold renewed her fear.

"Safety is relative at this point, my dear," Bae said quietly. Susan patted Carolyn's elbow.

"There's a mattress in back left." Ron gestured into the darkness. "I suggest you stay lying down while the bus is in motion. There will be a long stop not quite an hour in, at Glasgow, and another just south of Manchester in Knutsford. The next long stop will be London. Stay put! Someone will get you after the luggage has been unloaded and put you on the bus to Paris. You'll have to move fast in London: be prepared. We don't have nearly the length of surveillance cover as we can get here."

Bae stepped forward and gave Carolyn a hug. "Find whatever it is, girl."

Carolyn hugged her back. "Take care of Ellen."

"Don't worry, we will."

"Tell her—"

Bae clasped Carolyn's hands. "We know. You've said so many times." She smiled softly, sympathetically. "Ellen will understand why you had to go before she got here." She held up a hand to stop Carolyn's next words. "We'll tell her you love her, and we'll not let her worry." She squeezed Carolyn's hand. "You take care."

Ron said, "We're almost out of time." Susan ducked and crawled into the hold.

Carolyn bent down and scooted after. She turned back. "Thanks."

"Don't mention it," said Bae, with a sudden wild grin.

Carolyn found the mattress in the dim just before the door started closing. It was surprisingly firm and gave off a chemical, new-plastic scent. She crawled onto its springy surface, feeling bizarrely self-conscious of her shoes. She looked back at the silhouettes of Ron and Bae outside. Descending black eclipsed their forms, until the thin line of night light vanished with a clunk.

* * *

Carolyn woke as the surface underneath her moved, surprised she had drifted off with all the adrenaline rushing through her. Susan snored lightly beside her. She had not been the only one. But it had been a long day.

Susan stopped snoring. "We're on our way?"

"I presume," answered Carolyn. The silence stretched. "I never said thank you."

"For what?"

"Getting me to Uncle Keith." Carolyn blushed in the dark. "Trying to warn me."

"I didn't do the best job there," Susan said with a wry tone.

"Well …"

"I really thought you were in on it. Or at least knew a bit more. Had trouble processing the new info."

Was Susan taking her thanks as an opportunity to apologise for that bizarre meeting in the alley? "It's okay. I should have listened."

"Probably wouldn't have made a difference by that point."

"I suppose." Carolyn mentally reviewed the timeline. Could that have only been *last night*? Her life had changed and ended! so much since then. "If you hadn't followed me to the train station, what would have happened?" What if she had spoken to those men in the lecture theatre? Had they been planning her 'death' already? She swallowed. Had they been planning it to be more real than fake?

"I truly don't know," Susan said. "They tried to vanish me, but I don't have something they want."

Carolyn touched her father's ID tucked under her shirt. They both still wore them, it seeming the safest method of transport. "I have no idea what that is." Although the Schwarz Final Findings was looking less and less unlikely, as ridiculous as that was. "If I had the internet, I'd be reading my parents' papers, to get some idea." She made fists. "I'm kicking myself for deliberately avoiding their work, now."

"It's not as if you could have anticipated this." Susan' tone was comforting, reasonable.

"I suppose not." The mattress slid sideways; the bus must be making a turn. A sudden jerk in the other direction, and Carolyn found herself balancing on the edge of the mattress. She shifted closer to the centre. This bus ride was like her life, plunged into darkness and careening out of control. How could it have changed so fast? "I never read Tonya's paper." She sat up. "There were reference letters I should have done *today*! Students could miss out because I'm late …"

"Lie back down." Susan grasped her arm and pulled. "You're not late, Carolyn; as far as they're concerned, you're dead. You can't do anything about—"

"But—" Panic rose. And grant reports, and Frank's new findings: so many things undone. She could not just leave it all behind!

"You really can't do anything." She tugged at Carolyn again. Her voice was soothing.

Carolyn reluctantly lay back. "But—"

"You can deal with all that after this is over." A soft pat on her elbow. "For now—"

"I need to find whatever my parents didn't leave me." She scowled in the dark, panic turning to anger at the people who had done this. And just a tiny bit at her parents for failing to give her the thing everyone wanted. She tried to control her breathing.

"Back on the train you said something about a funny story about Bard bodies."

"Barr bodies," Carolyn corrected automatically. Susan was trying to distract her.

"What was it?"

The distraction was working. She found herself smiling. "It was the middle of the twentieth century, just a few years before Watson and Crick came out with their DNA structure. We actually knew quite a lot about genes back then, although people had trouble accepting it—DNA just seemed too simple, so they kept holding out for proteins to be the answer. But anyway, that's just to set the scene. Genetic science was pretty advanced at that point."

"Uh huh," said Susan, encouragingly.

Carolyn relaxed further, letting the story take hold. "So Murray Barr and his PhD student weren't involved in any of this. They were looking at neurons and trying to see if they changed physically after firing. They happened to be using cat neurons, and found something curious. About half of them had these little dark spots around the nucleus—where the genes are. They eventually figured out the half with them were from female cats.

"At first, they couldn't believe it. They though it must be some artefact. Surely, if you could tell a female neuron from a male neuron just by looking at it, someone would have already noticed!" She laughed. Sometimes the simplest things could go unnoticed in science. How many people had thrown out penicillin-contaminated petri plates before Fleming made his discovery? "But it turned out no one had. They found it wasn't confined to neurons, nor to cats. The dark spots were the Barr bodies—named by others, later—the extra X-chromosome."

"So you can tell a male cell from a female cell, just by looking?" Susan asked.

"You do have to stain it first, to visualise the chromosomes, but yes."

Susan moved, and Carolyn guessed she was looking at her arm again. "How weird."

*   *   *

London was a long period of waiting followed by a mad dash through the bus yard. Thankfully just long enough to visit the toilets: Carolyn had dreaded using the jugs Bae had given them for that purpose. Unfortunately, while Susan dozed through the ferry crossing, Carolyn discovered a hitherto unknown sea-sickness. She gripped the edge of the mattress and wished for the trip to be over.

Finally, they bumped onto the ground and took to the road again. Carolyn breathed deep, still queasy.

Movement next to her suggested Susan had woken up. "We need a plan for Paris."

"Uh …" moaned Carolyn. She did not feel well.

Susan ignored Carolyn's distress. "Vivcor Paris has been closed down for a few years now, but I suspect the site still exists. Given they were looking for the IDs—"

"Maybe." They did not know *what* they had been looking for. Mentioning IDs might have been fishing.

"True." Susan's tone was contemplative. After a moment of silence, she said with conviction, "But given that they think they could be useful, we can guess that the IDs might get us in."

"But in, where?" The lack of the internet was crippling. Carolyn had never gone anywhere without being able to look up directions.

"We have a few more options in Paris," said Susan. "The continent has less surveillance. I did an exchange there a few years ago; the police don't get data automatically from the subsids like we do. It'll take someone asking to actually find us."

That was something. "Although they think we've gone to Paris."

"That is unfortunate." The mattress wiggled, suggesting some gesture from Susan. "By now, they know we didn't get on the train we got tickets for. With luck, they're looking everywhere *but* Paris." They were hopeful words, although the tone did not quite match.

"But how do we get information?"

"There are ways," said Susan.

More police tech, Carolyn supposed. She closed her eyes. It was Saturday. She had 'died' last night. Uncle Keith and Bae would be on their way to get Ellen. They were not going to share anything with Nicholas, but would tell Ellen everything on a return trip via Grampian. Bae had her own tech to find dead zones.

At work everyone would believe the news stories. Who would supervise Frank's PhD? Did he even know his supervisor was supposedly dead, or was he in his office, prepping for Monday lab meeting?

*   *   *

Carolyn and Susan lay quietly as shadowed forms grabbed at and pulled off the luggage. There were a lot more bags on this crossing than the night bus from Stirling. Finally, the door clanked closed. The bus moved, spent some time jiggling forward and back, then was still. After another ten minutes or so in darkness, the door opened again.

Two silhouettes stood in the door's light. "C?" said a man's voice in a French accent.

"My name's—" Carolyn started.

"No names. Call us One and Two."

Susan crawled to the opening. One—Carolyn supposed the speaker was One—reached to help her out. Carolyn's trouser leg jammed on a strut, and she paused to work it free.

Two, another man, spoke as Susan stood up. "We've got a bunk for you in town, under the rail station. We'll show you the cam-free way in, and after that you're on your own."

"Understood," said Susan.

"This way." The men walked towards the front of the bus.

Carolyn, trousers finally free, scooted out and started trotting to keep up. They were in a large building where buses parked in seemingly endless rows. It was dimly lit by sunlight filtering in through high windows. She paused and looked behind her, wondering how the bus had gotten in, as more buses were parked directly behind the one they had left as well.

Two was slightly ahead. He stepped out past the front of another row of buses, and a dark shape ran into him from the side. People were everywhere, falling like rain. Carolyn dropped to the ground and rolled under a bus. A forest of feet trampled the space between the buses she had just occupied. Grunts sounded. She saw Susan pressed to the ground, struggling, then lifted out of sight. She could not tell which feet, if any, belonged to their guides.

Chapter Nine

Voices spoke, in French: *Il y a seulement une femme*; *Elle a la carte*; *Bien*.

The feet milled about, then left. Carolyn lay frozen, heart beating in her ears, for what seemed like ages afterward. A giant clunking sounded somewhere in the building, and distant French voices spoke and receded. She finally crawled out from under the bus.

She looked up. Where had they *come* from? She looked around, then cautiously walked past the front of the bus. There was no sign of the attackers, nor Susan or their guides.

She collapsed to the ground and leaned against the bus. Emotions warred within her: guilt that she had hidden while her companions were taken; terror that she was now *left alone*. Susan and her steadying influence—Susan and her police tech and her street-wise savvy—were gone. The far end of Bae's transport chain was gone as well.

She buried her face in her knees. Just yesterday morning she had given a lecture. How had the world changed so much? She wanted to give up, to make it all go away and go back to her old life. But that was not an option. Giving herself up to these mysterious violent Vandals was more frightening than doing nothing. She wished she had just hid back in Scotland; Bae's contacts surely could have squirreled her away into some off-grid existence. For the rest of her life. And what about Ellen? Would Ellen have come with her or been able to visit?

No. That was not acceptable. This needed to be sorted. She could not live in fear, and even more she could not let her daughter live in fear.

Plus, her parents may or may not have left behind something of such import that somebody was willing to go to extraordinary lengths to get it. Never mind fixing her life, she was filling with a growing curiosity. She *had* to know what it was. The scientist in her would not rest until the mystery was solved.

She hugged her knees. Fine thoughts, but she was alone in a bus depot in a foreign country, with no idea of how to get out without being found, much less find her parents' legendary Final Findings. And Susan who had come with her was now missing.

That provided a goal: she would get Susan back, then they would continue on. She stood. She had no idea where they had taken Susan, but if they were after the ID card that Susan had, they would likely go to where it worked. She needed to find the site of Vivcor Paris.

*   *   *

The building housing the buses seemed endless, but after walking a good ten minutes she found a wall. An hour later she had done a full circumference. There was no way out. She leaned against the wall, resisting the urge to collapse to her seat again.

The bus had entered somehow, and it clearly had not driven in. So, it must have come from below or above. She looked up at the skylights. She knelt and peered under the bus beside her.

A massive clunking sounded, nearby. She ran towards the sound. A bus was sinking into the ground. *Ah, they came from below.* Without stopping to think, she jumped on top of the bus as its roof reached floor level. The top was slippery. She flattened herself on her stomach and spread out her arms and legs. Perhaps this had not been the best idea.

At least the surface was wide. After a swaying ride downwards, French voices came from the front of the bus, then the engine fired up. She was going to slide off in the streets of Paris and die. She pushed herself to the back of the bus.

It was still dim wherever she was, darker than above, and her eyes had not adjusted. Hoping there was an exit at this level and that the bus had no rear windows, she lowered herself off the back of the bus and dropped. Her feet hit the hard ground; pain shot up her legs. But she was still standing.

Red lights on the back of the bus lit; a shrill *beep-beep-beep* filled the air. She fled as the bus backed up, running to a concrete wall festooned with boxes and cables. The bus left. She breathed a sigh of relief.

She was in some sort of tunnel, lit by means she could not discern. A chemical stench of batteries, with undertones of rubber and metal, surrounded her, giving a claustrophobic sense of heaviness. Forcing herself calm, she followed the wall until she came to a door, behind which was a set of stairs. She climbed them and found another door. It opened onto a streetside, lined with trees, on the far side of which was a river. She took a deep, relieved breath in the fresh breeze. She stood along a stone wall, interrupted at regular intervals with steel-bar covered arches and separated from a footpath by a few metres of carefully tended grass. She looked across at the water. Could that be the Seine?

She was conspicuous, standing beside the door. She strode out onto the path and walked purposefully. It was not long before she passed a public info booth. She stopped. Susan had said the subsids here did *not* funnel information straight to the police. Someone would have to be looking for her to ask for booth vid. She could wander aimlessly in Paris or try to get some information.

Although, someone knew they were here, as they had been ready for them. She kept walking. At the very least, she should use an info booth farther from the bus depot. She walked for about an hour, the area around her getting busier and more touristy. Eventually she came to the iconic glass pyramids that she recognised as the Louvre. An info booth here would have tons of queries.

A convenient booth stood near the entrance to the Louvre. She waited as a father with two toddlers hanging off his legs finished his search. She entered and chose to use the touchpad instead of speaking. She took a deep breath: in for a penny, in for a pound. She typed *Vivcor Paris*. The screen immediately showed a shiny white building nestled beside other shiny white buildings, across a grassy field. She had half expected nothing, from what Susan had said about it being gone. But perhaps Susan was wrong. Maybe it was still there, and she could slide inconspicuously into a stream of workers the next morning. The metro could get her within three kilometres of the place. She deleted her search from the screen and headed towards the metro entrance. It seemed too easy.

* * *

The three kilometres went past a pond and through a mostly wooded area. She had not realised there was so much nature so near the centre of Paris. She finally reached the grassy field. The buildings were not as shiny as in the photo. She continued along a road that curved around the edge of the field, not wanting to cross conspicuously to the building. As she hoped, it looped back to pass by the buildings, the first of which pressed right up against the trees.

The area was quiet, save for a few scattered bird calls. She stared at the building to her left. Autumn leaves piled in the corner of the recessed central doorway. The interior was dim. Were all these buildings abandoned?

She continued on. It appeared so. Someone or some bot must be mowing the field, but the buildings themselves exuded a sense of disrepair. Susan had been correct: the place was closed down. There would be no morning rush to fade into here. Still uncomfortable walking straight up to it, she passed into the alley between Vivcor and its neighbouring less-than-shiny white building.

About halfway down, a smooth door sat recessed into the wall, lacking handles. She pushed anyway, just in case, but it did not budge. She pulled her father's ID out from under her shirt. A subtle square patch to the left of the door looked promising: she waved it past. A beep sounded. For a few seconds, it appeared that was all that would happen. Then the door clunked and jerked open a few centimetres.

Her heart beat fast, sudden fear intruding. Were Susan's captors already here? What did Carolyn think she could *do*? She set her jaw, remembering her anger. These people had destroyed her life, when she would have happily shared whatever they wanted given a polite request. Susan had stuck with her and come to harm. She felt a sense of responsibility to her companion, who could have easily left her in Scotland rather than continuing on this chase.

She put her fingers in the crack around the door and hauled it open, then stepped in. The door had a horizontal handle on this side, and she pulled it shut behind her. The resulting clunk seemed like the sound of a door closing on more than just outside; it sounded like a door closing on her old life.

She stood in a dim corridor, muted sunlight filtering in from doorways further down. The floor was heavy with dust. No other footprints marred it; she was the first one through this door at least. She could not help but leave an obvious trace of her activity. She shrugged and walked forward. The diffuse sunlight came from laboratories lining the corridor on its left side. She peered in one.

Other than the dust, it looked like an average research lab. Pipettes stood in stands; a few lay on benches—some even still had tips attached. Labelled bottles crammed shelves. Bench spaces revealed personalities: some were meticulously organised, with large clear spaces; others had barely a square foot of working space with various bottles and tubes cluttering the surface. One of the meticulously clean spots had a single rack of microfuge tubes, caps open, a pipette with tip laying diagonally, and a crumpled tissue wipe beside it.

It was if everything had simply been put down in the middle of work. This place had not just been abandoned; it had been evacuated.

Chapter Ten

New fear spiked through Carolyn. What would cause a biotech lab to be evacuated, and was it still here? She backed away from the lab's open door. She should go back the way she came and leave whatever contamination might exist around her. But she headed deeper into the building instead. If her parents' secret was here, she had to know it.

She found a lift right before a set of fire doors, and a staircase directly past. The stairs were lit by small emergency lights inset along the edge of each riser. She chose the staircase. Something powered the door that let her in and the emergency lights, but whether a lift would work or not—or continue working if it began—was an unclarity she was not willing to risk.

One floor up she found a wide open space, with a smooth, dust-free floor. Cleaning bots must still be at work on this level. Windows to the outside formed two walls, the one towards the front of the building descending below floor level past what appeared to be a wide set of stairs. The other sides were lined with plaques, most of which appeared to represent patents. A bank of lifts sat directly beside the door she had entered, at the centre of which was a directory board. She scanned the board. Her eyes stopped at *Dr R Schwarz, Dr M Schwarz, Biogeneration, Laboratory 544.*

Her parents. She resisted an urge to clutch her father's ID card. Somehow this felt more direct contact than going through their boxes at Uncle Keith's: that had been material deliberately stored; this was a slice of their life, frozen in time.

She should not stand about in contemplation. Susan and her captors could be here, or on their way. Carolyn retreated back into the stairs and climbed. After four flights, she pushed the door open onto another corridor like on the ground floor. Light came in from what must be laboratories on the left; this floor was dust free. She walked cautiously down the corridor, looking behind at regular intervals. But it still seemed deserted.

She passed two fire doors before the lab numbers reached the 540's. Soft clunks came from the lab that would be one before her parents'. Adrenaline rushed, and she ducked into the nearest doorway, Laboratory 542. Like the labs below, it was full of benches abandoned in the midst of work. The floor continued to be clean, but the benches held a thick layer of dust. She peered around the door into the empty corridor.

More clunks and whirring sounded, getting louder. Her heart seemed to beat in her ears. She ducked back behind the door. The whirring changed pitch and came towards her, like something mechanical trundling along the corridor. A cleaning bot? She stepped back as the sound paused at the doorway.

The door pushed open, and a low, black cleaning bot entered the room. She stepped back further. The bot stopped, then spun to face her. She froze. Could it sense her? Did it have any defences? Surely not. But her mouth felt dry, and her palms sweated.

The bot moved again, turning away and running along the wall. She leaned against the bench to her right, then slowly lifted herself up to sit on it. Dust billowed, filling her nose and mouth with a stale, dry taste. She coughed. The bot spun back towards her again, but continued on after a moment. She hugged her knees to her chest and sat as still as she could as it methodically criss-crossed the room. She held her breath as it cleaned directly below her.

Finally, the bot left. The whirring changed pitch again, then muffled. It must be in the next lab. She allowed herself to relax. She slid off the bench and brushed dust from her trousers, leaving a showering of it on the clean floor. The bot would have more to clean on its next visit.

She returned to the corridor and walked quickly to Laboratory 544. She stepped inside. Her fingers tingled. She could barely believe she was standing in her parents' laboratory.

It was disturbingly familiar. Stacks of petri plates with thin amber sheets of what must have been dried-out agar media surrounded work areas. She stepped closer, finding the plates labelled things like *YPD* and *-Leu*. She stared. YPD, or Yeast-Peptone-Dextrose, was the standard growth media for yeast. -Leu would be media missing the amino acid leucine. Such media was used to select for successful transformation when doing genetic manipulations. You started with a yeast that was lacking the ability to make leucine, an amino acid essential for the organism to survive. You put in a bit of DNA—or *transformed* the cell—that had both whatever gene you wanted to add plus a gene to make leucine. That way only the cells that got your DNA would survive.

What were her parents doing studying yeast? Yeast was *her* organism. Were not they meant to have studied artificial genomes? She knew that back at the inception of that field yeast was a workhorse. Its ability to put together long sections of DNA and hold them in YACs, Yeast Artificial Chromosomes, enabled people to build genomes piecemeal and keep them in yeast cells until they were ready to be put together. But that stage had been passed long before Carolyn was supposedly designed. People were able to synthesise entire chromosomes, including attaching the proteins that were an integral part of holding the DNA in the chromosome structure, using faster artificial methods. There should be no reason for her parents to be using yeast.

She walked further into the lab. Like the first lab she had seen, it looked left in the lurch. A microdissection scope even had a dried-out agar plate in place, upside down over the microscopic needle that could be used to move individual cells. Old-fashioned paper lab books, with dark blue covers, lay scattered on benches. She picked one up and flipped to the last recorded page. It had the recipe for an agarose gel—used to separate DNA by size—and a list of samples with cryptic designations like *GH3-A-Day1*. Then the pen had scrawled across the paper.

The impossibility of her search hit her. How many people worked in this lab? How would she even know the secret research if she found it? She looked for more lab books. They had e-lab books back then, right? But if they were scarce, and expensive, perhaps only the most important details would get recorded there. She put down the lab book, only just then noticing that the benches here were not covered in dust like the other lab.

Before she could follow up on the implications of her observation, voices came from the corridor. Fear spiked through her. That was no bot. She dove and ducked under a lab bench, hiding in the space made between an incubator and a freezer.

"I told you, I'm not her." It was Susan's voice.

Relief flooded Carolyn. Susan was alive and okay enough to be talking. A tiny amount of the guilt that she held from hiding during the fight released. But she was hiding now.

Revealing her presence would not help matters. She examined her hiding place: power cords and pipes presented no ready weapons. Something thin and metallic glinted, shoved into the cooling coils on the back of the fridge. She reached for it, finding it the edge of a larger flat item. She tugged it out.

An e-lab notebook! She clutched it to her chest. That had not fallen; someone had hidden it there. It could be what everyone was looking for. She felt a wild impulse to jump up and yell out her find. She ruthlessly suppressed it.

Susan and her captors were coming closer. They stopped at the end of the lab bench bay, in front of a pair of dark computer monitors. They shoved Susan into a seat, which swung to face Carolyn. Carolyn could see them easily, meaning the men would be able to see her if they turned around. She held still, barely letting herself breathe. One of the captors lifted a device from the table and jabbed it into Susan's arm.

"Ow! Hey, what was—"

The man stared at the device, which looked something like a sleek, black ear thermometer. "She's right. No match at all."

"*Non? Mais elle avait la carte!*" said the other man in French. Carolyn gathered his meaning through his incredulous tone, though she could not understand his words.

The two men stared at Susan, and Carolyn's heart clenched. What would they do, now they knew she was not Carolyn? Carolyn took the risk to move, gently waving her arm. Susan's eyes widened. She had seen Carolyn!

The men continued to converse in French. Slowly, Susan made a fist with a thumb out and angled her thumb first at herself, then at the man to her right. Carolyn pointed to herself and to the man on Susan's left. Susan stuck three fingers out from her fist, then brought them back in.

She lifted them again, one at a time: one … two … three.

Carolyn launched herself at her target, clutching the e-lab notebook like a rugby ball. She was vaguely aware of Susan moving, faster than her, already scuffling with the other man. She hit her target in the stomach with her head. She stood up and caught his chin on her skull. That hurt!

But he was disoriented, and she aimed the heel of her hand at his nose, picking up a stance from a self-defence course she had taken some twenty years ago. He jumped back, avoiding her blow, but then Susan was on him, and before Carolyn could process what had happened, the man lay face down on the floor, unconscious.

"Thanks," said Susan.

Carolyn shrugged awkwardly. She had done barely more than distract one man. Susan had taken out both.

"The rest will be on their way." Susan spoke as she patted around the men's pockets. She pulled out two thin black cylinders from their clothing. "Is there another exit?"

"I don't know," Carolyn said. "This is as far as I'd gotten."

Susan moved to the door, then backed up quickly. "They're on the their way."

Carolyn and Susan retreated further into the lab. A door led into a small room with a desk, an office—her mother's or father's?—but there was no other exit.

Susan pushed on the window. "Help me!"

"We're on the fifth floor."

"You want to get captured?"

Carolyn leaned her weight against the window as Susan lifted the catch. It jerked outwards.

Susan put a leg over the sill and looked down. "There's a ledge." She slid out, then vanished below the level of the window.

Carolyn's breath came rapidly. French voices cried out in exclamation. They must have found their unconscious companions. She mimicked Susan's movements, swinging a leg over the window sill.

Chapter Eleven

Carolyn could not see the ledge her companion was standing on, but she slid to lay her chest on the sill and brought her other leg over. The e-lab notebook scraped against the sill, and she stopped to shove it in her backpack.

"What are you doing?"

"I found an elly-book. It might be important."

Susan's hands guided her legs, and she lowered herself. Her toes came to rest on something horizontal. She still held the window sill, at about eye height.

She looked down. The surface barely extended to her heels. "You call this a ledge?"

Carolyn pressed closer to the building. The stonework was heavily ridged, with some sort of carving. It was one of these protuberances that the shorter Susan was holding. They were at what must be the back of the building, with an empty car park below them and woods behind. She had never had a fear of heights, but had to admit this did not make her precisely comfortable.

"Ledge enough that we didn't fall, did we?" Susan looked up. "They've found their friends. It's only a matter of time before they came the same way we did. Don't suppose you could swing the window closed?"

Carolyn reached up with one hand. She grabbed the sill again, unbalanced. "I'm not sure."

"It'll give us more time."

She slid her knapsack off and handed it to Susan. She tried again, this time reaching the window. She pushed it back in place. She had not intended to push it so hard, but it clicked shut. They were truly stuck out here now.

"Come on, there should be a drainpipe somewhere." Susan sidled away. She went considerably faster than Carolyn, who moved one careful hand at a time to a new piece of carving. The stonework was cold and gritty under her fingers. How strongly were these bits attached, anyway?

Susan waited for her at the corner of the building, where a drainpipe curved over the carvings and ledges. "Okay, it's secured every few feet by bolts. You can use them as footholds. Watch me."

Susan monkeyed down the drainpipe, holding the pipe in both hands and placing her feet on something Carolyn could barely see. Carolyn waited until

Susan was a floor below, then cautiously approached the pipe. She held on to it as Susan had. It wiggled with Susan's climb. With visions of the pipe pulling away from the building from their joint weight, she let a foot feel for these 'bolts' Susan claimed existed.

There was something there, a bare nub, but enough to give some traction. She followed Susan down, unabashedly stopping each floor to rest on what now seemed a quite substantial ledge. Finally, she reached the bottom.

Susan grinned. "Good job."

Carolyn's chest swelled at Susan's praise. She looked up. Had she really come down that? Yesterday morning in the lecture theatre seemed a lifetime away.

* * *

Susan strode off through the woods. Carolyn had to walk quickly to keep up, despite her longer stride.

Susan finally stopped checking behind them. "So, what was this you found?"

Carolyn gestured towards the knapsack Susan still held. "An elly-book. I assume it's an e-lab notebook. It was jammed behind a fridge. It looked like it was something someone tried to hide." She recalled her despair at the sight of all the lab books. "They were pretty new and expensive back then. I'm hoping that it stored the most important stuff."

Susan patted the knapsack. "They're probably still searching for us in the building, but they'll widen the net eventually. We need to get somewhere safer before we can examine it."

Carolyn reached a hand towards the knapsack. "But it might have what everyone is looking for."

Susan patted her outstretched hand. "More reason to get safe with it. How did you get here?"

"I took the metro." She felt self-conscious as Susan frowned, scanning their surroundings. "I didn't know what else to do—"

"No, that's fine. You did fine. But we need a more private way back." She stopped and put a hand on a tree. "Which metro stop?"

"Chateau de Something. It was the end of the line."

"Château de Vincennes?"

"Could be."

"Okay." Susan headed off at a slight angle to her previous path.

The forest floor crunched underfoot, autumn leaves crackling and small twigs providing occasional sharp snaps. Carolyn found the walking difficult. The ground was uneven; they wandered around trees and occasionally ducked under deadfall. But Susan strode forward as confidently as if she were walking on pavement.

"What was that thing they did to you?" Carolyn asked.

"What?" Susan said.

"The thing they poked you with. It hurt."

"Not entirely sure. A genetic test, I think, from what they said on the way. I kept telling them I wasn't you, but the one fellow said they could just use the test to be sure, and then something about opening it for them."

"Opening what?"

"No idea."

A sound like a car engine grew and receded. Carolyn asked, "Did you hear that?"

"Means we're going the right way. There should be a road, right up ..." Susan fell silent as the space between the trees brightened ahead.

The woods ended at a small, two lane road. Dirt paths edged the sides, appearing more to be grass trampled away by pedestrians than any planned walkway.

"Don't suppose you got pen and paper in this bag of yours?" Susan flipped the knapsack from one shoulder and twirled it around her front.

"I don't think—"

"Ahah!" Susan pulled out a piece of cardboard and a thick marker. "Your aunt knew her stuff." She scrawled *Paris* on the cardboard.

"We're going to hitchhike? Isn't that dangerous?" Carolyn shrunk back under Susan's plainly sarcastic gaze.

"Dangerous for us now is the grid—public transport, main streets, large stores—anywhere they might have surveillance. Not individuals."

Susan held up the sign. The road was busier than Carolyn was used to; she had heard people used more automobiles on the continent. Only four cars passed before one pulled alongside them. Carolyn hung back while Susan leaned in and conversed with the driver in French.

Susan opened the passenger doors front and back. She gestured Carolyn to the back door. "Get in."

Carolyn obeyed, and Susan got in the front. Susan continued speaking to the driver, who was a middle-aged woman. She appeared to be piloting the car herself, though the dashboard suggested it had self-drive capabilities.

It was not long before they entered a more city-like area. As the traffic got heavier, the woman moved to flip a switch on the dash.

Susan reached a hand out to stop her and leaned forward. "*C'est assez près.*"

The woman responded in rapid-fire French, to which Susan responded, "*Non, non. C'est bien,*" while patting the driver's shoulder. The woman pulled to the side of the street and let them out.

"Where are we going?" asked Carolyn.

"Our guides mentioned a camera-free path into the rail station. I'm hoping it actually exists." Susan headed off down the street.

"Under the rail station," Carolyn said.

"What?"

"Under. They said under. And—what happened to them?" Carolyn felt odd, realising she had not thought of One and Two since she had seen Susan. But they had been so strange, so impersonal, even foregoing names. It was easy to forget them as people. She wondered if this underground movement was not losing more of their humanity with their hiding than they claimed mainstream life was taking from everyone with corporate control and surveillance.

Susan paused, looking back at Carolyn. "They broke free. I …" She turned and kept walking. "I'm not entirely sure I trust them. They seemed to get free too easily."

Carolyn strode briskly to catch up. "Do you think they're with the Vandals?"

"I'm not sure those were the Vandals." Susan turned down a narrow street. "I fear there may be more than one player in this."

"Why?"

"The people who came to your uncle's were looking for prox cards, like the ones we found." She glanced at Carolyn. "Did that let you in the building?"

"Yes," said Carolyn.

"The people who grabbed me already had access."

"But they wanted your card." She didn't know French, but surely *carte* was a cognate.

"It seemed more an identifier than anything." She patted her chest. "I still have it." Susan leaned against the grey stone of the building beside them as they approached a larger street. "I'm looking for some kind of pedestrian subway, or …" She stood up. "See that square brick building there? The small one?"

Carolyn squinted. "By the intersection? With all the green?"

"Yes. It's got an anticorporate glyph. That must be it."

"A glyph?"

"That graffiti—the 3D red box to the right of the door." Susan frowned and looked around. "Okay, follow me. Look casual."

Look casual? Carolyn had no idea how to do that. But she walked after Susan, trying to avoid rubbing her hands together nervously as they waited for crossing signals.

They reached the small brick building placed oddly in the centre of the intersection. Susan stuck out a hand and leaned against the door. It opened. She stepped inside quickly and pulled Carolyn after her with a sudden hand on her elbow.

The door shut behind them. Inside was dark. Susan's hand left Carolyn's elbow, and Carolyn froze, disoriented. Grunts sounded, from more than one person. Carolyn reversed until her back pressed against a cold wall.

Chapter Twelve

A faint light flared, illuminating a scowling woman's face. "*Qui êtes-vous?*"

Susan spoke rapidly in French, and the woman's scowl gradually relaxed. As Carolyn's eyes adjusted to the dark, she could make out two other figures behind the woman. Susan gestured towards Carolyn, and one of the other figures nodded.

"This way." He started down what turned out to be a metal spiral stair set into the floor. The air warmed as they descended, until Carolyn opened her coat. They reached the bottom, where a maze of white, smooth-walled tunnels extended out like some underground spider web. Unlike near the bus depot, the air smelled fresh; they must have good ventilation. Their guide did not speak for all of the what must have been twenty minutes or so until he stopped in a short corridor with six doors off of it. He gestured to one door, handed Susan something, and left.

Susan pressed the something into the door and fiddled a moment—it appeared she was using an actual physical key. She must have been: she pushed the door open. As the door opened, a square panel in the ceiling flickered into life, lighting the room below. Inside were two cots, with thin pillow-like objects and rough blankets, pushed up against the right and left walls. Between them sat a small wooden table with two chairs.

"Whatever One and Two's dubious loyalties, they were honest about the bunk." Susan sat on one cot and swung the knapsack around to the floor.

Carolyn closed the door behind them. "How many people do you suppose are down here? Are they all hiding from the corporations?"

Susan shrugged. "Most probably aren't hiding like us. They just want to be off-grid. There are a lot of small communities like this. A few teens and new adults, but mostly elderly and families. The police know about them. As long as they don't make trouble, we ignore them."

"I never knew." Carolyn sat down across from Susan. So much underfoot—literally, at least in Paris—that she had been completely unaware of. Her gaze wandered to the knapsack. Susan was clearly thinking the same thing, as she reached in and pulled out the elly-book.

Carolyn's heart thumped. She leaned forward and took it, her hand shaking slightly. This could be the answer. The elly-book was black and slim. She felt along the edge and found the power switch. The screen burst to life. It felt all too familiar. Queasiness fluttered in her stomach. A welcome icon appeared, of an eOS only a few versions out of date. Disappointed, she blew out held breath. "This is a modern elly-book." It had been a coincidence too good to be true, that she had hidden exactly where someone thirty years ago had stashed her parents' secret research. How could she have not realised that?

"Well, what's on it?" Susan leaned forward, sounding still interested.

It had been hidden. Perhaps it was useful. The main screen filled with picture album icons. Or perhaps it just stored someone's photos. She tapped on one. A lined page filled the screen, covered in scrawled blue writing: lab notes! She bent over the screen and flipped through photos. Excitement set her knees jiggling. "Each album is a lab book!"

She looked across at Susan, grinning wildly. All those lab books that had discouraged her: someone had gotten there earlier and photographed them. They must have been interrupted and hid just where Carolyn had, as one of the few obvious such places in the lab. Not so much a coincidence after all. Perhaps the elly-book did hold the answers.

*   *   *

Carolyn accepted the sandwich from Susan. She had been reading the lab books for hours now, during which time Susan had investigated the obscure black cylinders she had taken from the Not-Vandals, gone exploring, and finally returned with food.

"It doesn't make any sense," Carolyn said. Susan gestured with her own sandwich, her mouth full of food, and Carolyn continued, "This research. It's all … old news. Even back then. I don't understand what they were doing."

"What do you mean?"

Carolyn frowned, trying to think of how to explain. "I've gone through over thirty books now, covering at least two years. They're doing experiments and recording results, but it's more like a teaching exercise instead of real research. Except … it's not presented that way. It's written out just like you'd expect for new work, but this stuff was well known decades earlier."

"Could they not have realised? Be repeating research?"

Carolyn shook her head. "It's just so basic; that's impossible. They were synthesising the violacein pathway, which anyone in synthetic biology knows was one of the go-to proof-of-principles back when the field was beginning. It has excellent tuning properties and makes different coloured pigments depending on flux through different branches of the pathway. Perfect to show that your tools could modify gene expression exactly as you planned. But here …" She shook her head again, this time in confusion. "It's like they're finding it for the first time. Measuring everything that everybody knew already. I don't get it."

"Were they testing something? Perhaps some control mechanism?" Susan waved her sandwich-free arm vaguely.

Carolyn could nearly feel the *this is your bit* thought emanating off Susan. Susan had navigated them through the shadowy world of off-grid existence. Carolyn was meant to understand the science. "They were doing a few things, like tying the output to some mitochondrial functions, but there was no real science behind it. More like making some pretty yeast." They would have been pretty, changing colours in response to carbon substrate, some oscillating in various Turing patterns—spots and stripes, formed through processes similar to those that made pelt patterns on animals. It was more like art or … "Like it was for show!"

Her heart jumped into her throat, and she tapped on album after album, trying to find earlier lab books. She had been going forward, looking for the development of the research. But it had gotten more and more artificial over time. "They found something," she said, sure of it now. Despite all the interest, she had still harboured doubt that secret findings existed. Her parents had been great fabricators. But here, they had been fabricating show on top of nothing research. "Something they didn't want anyone to know, even their bosses. They spent years making pretty yeast, trying to draw people away from whatever it was …"

She bent over the images. Farther back, mitochondria featured more and more. Her parents had traced a path from mitochondria—*that's my subject*, a tiny piece of her protested—to something impressive-looking but meaningless. But where had they started?

The earliest lab book was three and a half years before the latest, and it showed the very first steps of tying the violacein pathway to mitochondrial cycles. Whatever the secret was, it was in the mitochondria. Mitochondria, what Carolyn had made a research career on, in attempt to avoid following in her parents' footsteps. A strange sense of jealous protectiveness gave way to mirth. She covered her mouth as laughter bubbled out.

"What?" Susan asked.

"It was in the mitochondria!"

Susan stared, clearly not comprehending.

"Sorry." Carolyn got herself back under control. "It's just that I decided to study mitochondria to be different from my parents. It looks like I followed them anyway."

"So, what's the secret?"

"I still don't know. But I think this book is where they started hiding whatever it was." She tapped the screen. "It's dated …" She zoomed in on the date. "Just three months after my birthday. I was a baby, living with them." She stared in wonder at the date. Of course, she had been with them for all these books, but so close to her birthdate … She chewed her lip. She had not been born in Paris. She straightened, the pieces coming together. "I was born in Bucharest. They moved when I was a baby, to here, and started this … distracting research. Whatever they found isn't here. It has to be where they were before." She crossed her arms. "Is there a Vivcor Bucharest?"

"There is." Susan leaned back. "Or at least there was, a few months ago when everything started going wrong for them."

Carolyn grinned. "Let's find it."

Chapter Thirteen

Carolyn and Susan stepped out of the autocab along the side of one of Bucharest's highways. The morning was chill, and a stiff, cold breeze tugged at Carolyn's hair. Cars zipped past, most driverless autolounges, with passengers facing inwards towards each other. A few personal drivers with more traditional layout passed, but none appeared to be currently piloted by their occupants. Carolyn could see why: the whizzing traffic of the morning commute was dizzying, more suited to electronic brains than human.

A grassy hill to their right rose approximately the height of a two-story building. Paris's anticorporate hideaway was underground, in a maze of space-age tunnels made for some purpose Carolyn had never asked. Bucharest's hideaway was meant to be here, in the middle of the city, on the other side of that hill. She had trouble believing it. Perhaps the contacts Susan found were in league with whoever the Not-Vandals group was and had sent them on a wild goose chase. Or into a trap.

Susan looked up the hill, looked at Carolyn, and shrugged. She started up the slope at an angle, and Carolyn followed. Susan stopped abruptly on the summit. Carolyn joined her and stopped as well.

Before them stretched the strangest sight Carolyn had yet seen: it appeared to be a huge concrete bowl with a marsh inside. Tannish sides started vertical at the edges and curved down to vanish underneath tall, waving grasses. Streams and ponds dotted the area, as well as small hill-like bumps and stands of trees. A flight of birds took off from the nearest pond and flew, chirping, to another. A cuckoo called from somewhere farther out. "What in the world?"

"You'll know it when you see it," Susan said. "That's what they said. I guess this is it."

"But it's *huge*." The bowl must have been two kilometres across. Where would the hideaway be?

"We're meant to find the centre." Susan knelt on the edge of the nearest wall, then turned and dropped down, to land on the start of the curve about two metres below. Carolyn did the same. A considerably shorter drop than their last vertical descent, at least. Although the slope must have descended ten more metres, until they reached the edge of the grass.

A worn footpath was situated a few metres in from the edge and appeared to circle the perimeter. They made their way to the path and followed it until another peeled off towards the interior, on a raised ridge between two ponds. A cuckoo called again, then burst across the path in a whir of grey feathers. Carolyn gripped her elbows. She found herself in yet another strange world, something barely imagined those sparse few days ago when she had been just a biology lecturer with a medium-sized lab. Did her parents really have some super-secret research, so amazing that people were looking for it decades later? Perhaps her birthplace would hold the answer.

*   *   *

Their guide, a scruffy-haired teenage boy who intercepted them partway into the marsh, left them alone to settle at their bunk: a simple platform tent. A raised wooden floor had triangular frames at each end with a single pole connecting their points and a canvas tarp draped over. The outside of the canvas was covered in camouflage-like netting. But other than having been nearly impossible to find had they not been following their teenage guide, it did not in any way hide the structure. The camouflage must have been for aerial views.

There were a good ten cots in the tent, two with neatly folding blankets, and at the far corner, one with a rumpled pile of blankets and clothes. Carolyn sat on one of the cots clearly meant for them. "Wonder who else is here?"

"You don't ask," said Susan, "and they won't ask us."

It was a strange world. But they were alone for the moment. Carolyn pulled out the elly-book from her parents' lab. She had scoured it further yesterday on the day-long train journey from Paris, but had not had a chance to talk. Susan's devices—newly obtained from the Paris group—were unable to find any surveillance-free areas on their train. While the feeds would not have been automatically analysed, Susan said speech could trigger further scrutiny. So they had obscured their faces and remained mostly silent until arriving in Bucharest early this morning.

"They were definitely studying mitochondria," Carolyn said. It was probably less of a revelation to Susan than Carolyn. How had her parents gotten there from synthetic genomes? Although, she supposed, they never had really *done* synthetic genomes. "But everything here is a smokescreen. Most of it is measuring flux and spectral properties of one of the most well-studied biosynthetic pathways, things everyone knew already. But early on, there's some real stuff; they were tying the violacein pathway to different types of respiration, which at least is doing something, and they were measuring efficiency of the electron transport chain for some reason."

"Uh ..." Susan spread her hands.

"I'm not making sense, am I?"

"What do I need to know?"

Carolyn looked down at the elly-book. "I'm not sure. I'm thinking the earliest stuff would be closest to their secret. If they were trying to keep a line of research secret, they couldn't just drop it completely. That would bring people to ask about it—but if they just moved smoothly into something else, it would not be so obvious. So respiration and the electron transport chain must be related to what they were doing before."

"I still don't know what that is."

Carolyn couldn't explain respiration without writing on something. She paged to the end of the elly-book, but couldn't bring herself to write on it. But she had nearly forgotten: she had another—that elly-book she had grabbed on campus and never had a chance to return. She reached into her bag and found it far in the bottom. She pulled it out.

"You've got a second elly-book?" Susan straightened abruptly from her relaxed slouch.

"I found it the night you ... spoke to me." Carolyn paused, uncomfortable. She still did not understand fully what had happened there. But Susan's life had just been turned upside down, as hers had at her Uncle's place. She had been a bit wild, herself. "It looks like somebody's e-lab notebook. I was going to return it, but then those men in the lecture theatre …"

"It's not yours? You've had it with you *this whole time?*" Susan's voice rose with twinges of panic.

Sudden fear, and embarrassment, surged through Carolyn. She sucked in breath. "Can they trace them? I thought they were off-net." She suddenly remembered Ellen's use of hers; they were *not* off-net.

"They've got location services! Whoever lost it could find it—and us—instantly." Susan's eyes were looking wild again, reminiscent of that night in the alley.

Oddly, Carolyn felt calm. "But no one knows *I* have it. I don't even know whose it is." She swiped it open, expecting to find a passcode screen from which she could access the scribble-pad, but instead two grey cats covered in icons glowed out at her. She tapped the upper left icon. Scrawled notes from her second-to-last lecture, on telomerase, appeared. "It looks like it belongs to a student." She might even be able to guess who, as so few had them. Perhaps the woman with straight brown hair who sat in the second row; she was usually tapping on an elly-book. She remembered her coming forward, hugging an old tablet, in that last lecture Friday. Had she been planning to ask about her lost elly-book?

Susan grabbed the elly-book from her hands. She zoomed through into the settings, then slumped. "It was sited, but last Friday morning, at what I assume is your place. No one's tried to site it since." She looked up, biting her lip. "If it's a student's, they might try again, if they need it for notes this week."

Carolyn stared at it. She had run off with someone's notes—from her class, and other classes too. She was 'dead'. She could not return it: not before she figured out what was going on; not before she had fixed things so that Ellen was safe. Would that be before exams? She hoped the student had a backup of these notes.

"Let me see if I can … oh, crap!" Susan tapped furiously, then leaned back and closed her eyes. "They sited it. It's blocked now, but I was too slow. Someone knows where we are."

Chapter Fourteen

"Will the student be able to access her notes?" Carolyn asked, mildly hopeful. They were used for e-lab books precisely because they were not meant to have external data connections, but given what she had seen on Ellen's elly-book, perhaps things had moved on.

"That's not the issue! Someone knows we're in Bucharest."

"A student knows her elly-book is in Bucharest. That's not the same." Carolyn felt guilty about the notes. A dead lecturer could not be expected to write reference letters and provide grant reports and all the other things she had left in the lurch. But dead lecturers also did not steal people's study notes.

"She knows it was at your place Friday morning."

"She wouldn't know where I …"

"How long is it before she goes to the police about her stolen elly-book? *They'll* recognise your apartment."

Oh. "And so will the Vandals."

Susan frowned. "If someone …" She threw up her hands. "Yes, they will. I hate that they're so connected to the police, but it has to be true."

"But we have two elly-books! We can tunnel to the web, now."

"What?"

Did she know something Susan didn't? "The kids are all doing it. I don't really understand, but my daughter …" She had managed to avoid thinking about Ellen too much. Probably on purpose. But she would sort everything out and Ellen would be safe. She shivered her shoulders. "Anyway, my daughter and her friends would chat. She said it was a tunnel, you needed two ends but once you had that, you could use the connection as some kind of third base to connect anonymously wherever you wanted." She took the elly-book from Susan and paged through the icons. "Here! This is it." She passed it back.

Susan frowned. "That's one of the kids' apps, for sharing photos. It shouldn't work on an elly-book." She tapped the icon. A screen appeared, showing photos slowly circling around a central point. "Maybe this is some kind of storage version?" She tilted her head. "That'd be a hack; the app is specifically meant to be ephemeral; photos only exist as long as it takes to transmit and view."

"No. That's not what it looked like." Carolyn took the elly-book and watched the photos pass. The brown-haired woman and several other students she recognised precessed around the screen. She filled with sudden sadness; she missed her students. She missed Ellen. She squared her shoulders. She had to think about something else. A photo caught her eye—unlike the others, there were no people. It showed a white cloud-like swirl on a blue background.

"That's it!" She pointed to the photo. "That's the screen Ellen had!" She reached to tap the image.

Susan grasped her hand. "We don't know how to use it, and we've already been pinged in the hideaway. We should get into town and explore there."

*   *   *

They sat in a park full of sculptures, near the tallest one—a sharp obelisk piercing a basket-like oval, red drips running down it like blood. Bucharest was a mishmash of so many things, it was hard to process. Ornate medieval buildings squished side-to-side with twentieth-century communist cubes. Shiny modern buildings slithered between them like veins of mica criss-crossing the city. There were sculptures *everywhere*, concentrated in this park in the centre of the city.

Carolyn held the student's elly-book, and Susan held the one with the albums of lab books. A few words with their teenage guide at lunch time, assisted by translating apps from Susan's Paris-provided tech, had produced a memory stick with which they copied the photo/tunnelling app to the discovered elly-book. Carolyn took a breath and tapped the tunnel icon. Susan did the same.

Carolyn leaned over curiously, wondering what would appear on Susan's fresh version. It had photos, of people, but not Carolyn's students. They were a strange mish-mash of families on beaches, young adults in offices, and people striding down the street. They looked almost like advertisements, but missing the slogans and something to sell. Stock photos?

"There!" Susan pointed at Carolyn's elly-book.

She looked back, just in time to see the cloud-swirl photo vanish off the bottom. "Hopefully it'll come back."

It appeared on Susan's before it reappeared on Carolyn's, and Susan tapped it. The white swirl filled the screen, then flashed through the other photos rapidly, each visible no more than a quarter second. Susan slumped. "It's just a large view."

The swirl reappeared on Carolyn's, and she tapped it as well. Her screen mimicked Susan's. Disappointment filled her. Maybe it was the wrong app. She could have sworn this was the one Ellen had been using, though. She watched both screens. Hers had more photos, like Susan's but also with the brown-haired woman's personal pictures. *More* photos. Why would hers have the stock-like photos at all?

"I'm not so sure." Carolyn concentrated on her screen. The stock-like photos were there far more often than they should be, in relation to their numbers compared to the photos she had seen on the initial screen. They weren't exactly the stock photos, either. They had ... words? No, a single or letter or number, each, somewhere on the image. *There!* That wasn't a stock photo of a beach sky, that was the cloud swirl again. She waited, and after a short while it reappeared. She stabbed, but it had gone, freezing the screen on a picture of three students at a table. She wasn't sure how to make the app go again; no amount of swiping, tapping or tilting seemed to work. She eventually quit and restarted the whole process. In the meantime, Susan echoed her mistake.

Several tries later, Carolyn was the one who first caught the cloud-image. It stabilised, and a blank box appeared in the centre.

That was what she had seen on Ellen's screen. She grinned wildly at Susan. "I think I need some number from yours. Ellen said you needed a code to connect."

Susan sucked in breath. "Where? There's no control panel or anything."

"Look in the images? I saw letters and numbers in mine."

"Yes! I see them." Susan waited, then spoke intermittently. "Three ... Six ... K ... Four ... Three ... Six ... K ... I think it's repeating."

Carolyn typed in '36K4'; it filled the box. The screen flashed white and the app quit. "That didn't work."

"Try again; I couldn't tell where it started."

Carolyn restarted the app. "Let's gather my code, from the start. Maybe that will help." They collected the string C51m from Carolyn's images, while Susan finally captured the cloud-screen to type it in. Susan's screen swirled in the same way Carolyn had seen Ellen's elly-book do, but nothing happened on Carolyn's. Susan's screen continued to swirl.

"Maybe you need to be at the start screen," Carolyn said, contemplatively. She waited for the cloud-image and tapped it. She was getting better at this. It filled the screen, but instead of a blank box, the centre held a message: *Bridge 6K43 requested. Accept? Deny?*

They did not look like buttons, but Carolyn cautiously tapped the word 'Accept'. Her screen swirled like Susan's, then both stabilised with a divided screen. The top of Carolyn's had 'Bridge 6K43 formed' in a text-bubble; the bottom was black with a white blinking cursor like a terminal window.

"This bit here is us chatting," Carolyn said, typing 'Hi', which appeared with a delay on Susan's top half. "I think the bottom is the web." She sucked on her lower lip. "It looks like some sort of text interface." She did a lot of bioinformatics using terminal windows, where you would set up batch scripts to query remote databases with your results. She typed the character *?* into the window, a standard way to access help from command-line programs. The screen filled with brightly coloured text on the black background, headed by *Lynx Help Page (1 of 3)*.

Carolyn had heard of Lynx, back in a technological history class; it was an early, text-based web-browser. They liked to name things after animals back in that era. There was something else named after moles or similar. She scrolled through the help. "Ah, you just type the address." She bit her lip. "Do you know any addresses?" Carolyn knew address of bioinformatics databases but not much else; she never had cause to use an actual text address otherwise.

Susan was staring at Carolyn's screen, mouth parted. She lifted her head to look at Carolyn. "It's the antweb. It's got to be!" She returned to staring at the screen. "In a *kids' app*?"

"The what?" asked Carolyn.

"The antweb."

"It says Lynx …" She got the sense Susan was talking about something else entirely, and not animals.

"I've heard rumours. The tech guys know more. The anticorporate crowd are meant to have some other web—deeper than the darkweb, even. But I never heard it even suggested it might be on an *elly-book*."

"Darkweb?" Carolyn wasn't following Susan at all. *Darkweb* sounded ominous. "This is just kids' stuff. Ellen has it. My student has it." She gestured over the elly-book she was using. Wind whistling through the sculpture above her blended with traffic noises from streets around, heightening a sense of the surreal. Ellen's words echoed in her mind: *There's them that use it for anticorp stuff* and *Aunt Bae said* … The reference to Bae had passed her by at the time, doing little but making the app seem innocuous via the implicit blessing of an elderly relative. But that was before Carolyn knew that Bae was an anticorporate radical.

Chapter Fifteen

Carolyn raised her head from the sink and looked into the mirror. The washroom in the marsh centre was draffy and cold, but at least the water was warm. Last night, the text-based browser—whether or not it was the antweb—had allowed her to very slowly download photographs and schematics of Vivcor Bucharest's headquarters. There were actually four sites in the city, but they hoped any records would be kept centrally. They were banking on her mother's ID card to get Carolyn in: she and her mother looked similar, down to both wearing their hair in a thick poof of curls, save for Carolyn's being a few shades lighter.

She squinted at her new hair colour. It looked completely black, rather than the deep brown she had been going for.

"It'll lighten." Susan stepped back, her hands covered in loose plastic gloves lined with hair dye.

"I hope so." There were so many unknowns—perhaps Vivcor did not use the same style cards anymore; perhaps people tapped in rather than just walked past a guard; perhaps Vivcor housed the Vandals and trying to enter on her mother's card would get Carolyn captured. But that last was unlikely—the Vandals had been looking for the ID cards, and surely Vivcor could access their own sites. As had the French Not-Vandals. Were they Vivcor?

Carolyn let Susan finish wrapping up her hair to set. They wound their way back to the platform tent. A small dog, brown on top with white belly and chest, slid from the tall marsh grasses and followed them. Carolyn sat on her cot cross-legged, facing Susan.

"Let's go over it again," Susan said. "You enter through the front door."

Carolyn took a breath, recalling the photos Susan had found of Vivcor's main entrance. "I bear left and take the first door past the toilets." It had been painted blue. "I take the first right, and the first right again until I've nearly doubled back to the lobby. Directly before the lobby, a door on my right should be the stairs. I go down the stairs and turn ..." She could visualise which way she wanted go—towards underneath the lobby—but not connect it to a direction.

"To the right," said Susan.

"Are you sure?"

"The stairs will turn you around."

"Okay." That almost made sense. "Then I deal with what I find." Records were supposed to be held underneath the lobby. They had no idea if someone worked the room, if it was locked, or what, but at the very least this would be a fact-finding mission. "You know, I have no idea why my parents left Bucharest in the first place. I wonder if this research had something to do with it."

Susan held a hand out to the dog, who cocked his head. "Maybe. You said it looked like they were trying to change direction. Perhaps that would be easier somewhere else."

"Maybe we'll know soon." Carolyn tapped her wrapped head. "Is it time to remove this yet?"

"Just about." Susan stood, and Carolyn followed her back to the washroom.

* * *

Carolyn slid into the wave of people entering Vivcor Bucharest Headquarters. She held herself upright and felt prickles along her back. *Relax*, she told herself. She did not want to look stiff going in.

People streamed past guards who did little more than cast lazy eyes across the ID cards hanging on chests. A few people tugged theirs out of coats, but Carolyn had opened her jacket so the card was visible. She walked in.

She forced herself to keep moving. This was meant to be a familiar view, not something strangely awkward having seen it only in a still image … which was clearly at least one redecoration ago. There was a giant art fixture—some kind of spiral, DNA?—smack in the middle of what had been open space. She bore left, but could not yet see the toilets.

There were no toilets nor door to a corridor. The redecoration had been larger than just art. She pushed from her mind the worry that the records room had also been moved and continued through a door at the far end of the lobby. She just needed to find her way down and back underneath where she had entered.

The corridor still had a lot of traffic. She moved with the flow, watching as people peeled off left and right. She followed a chatting couple through a door to the left: another corridor. They stopped and went into a room to the right; she kept going. She took the first non-room left she found; she did so again.

She should be nearly back to the lobby, but if she had not gotten turned around there could be no door, as she had not passed one before. Yet there was, at the end of the corridor. She hesitated, then went through. It was a staircase. Finally!

She went down. It folded back upon itself twice and took one blunt right. By the time she reached the bottom, she was completely turned around. She remembered Susan saying 'turn right' and went that way; if they had only changed the ground floor, maybe this was the same staircase.

The basement corridor gave the sense of being poorly lit, yet it had just as many glowing ceiling tiles as the floor above. Perhaps the knowledge that she was underground did it. She kept going far longer than she should. They must have moved the records room.

"*Hei!*" a voice called from behind her.

Carolyn turned, heart thumping.

"*Cine ești tu?*" A young man in pale blue overalls approached.

Chapter Sixteen

Hey, who are you? Carolyn mentally translated—surely it was that, or something similar. She boldly held up her mother's ID, letting her hand half-obscure the name. "Intern." Romanian was a romance language, wasn't it? Intern must be somewhat cognate. "Records. Archives." One of those should work. She didn't relish the idea of tapping out a conversation on the Paris phone. They expected a multi-national like Vivcor to use English as their main language; but perhaps not the maintenance staff.

"Records are not here," the overall-clad man said in heavily accented English. "You are the fifth one this week."

Relief flooded her; they could converse. Although was that a boon? She needed to take the initiative, before he asked more about her. "Where are they?"

"Moved."

"To?"

He shrugged. "Who knows. Maybe burned. Will any of us have a job next week?"

It did appear things were as bad at Vivcor as rumours said. She had no idea how to negotiate this conversation. "Well, um, thanks anyway." She backed up, turned, and headed toward the stairwell. Her back prickled until she reached it. She took a few big steps up the stairs, past the first turn, then froze. She remained tense until she heard the doors thump closed.

She sat on the stairs. She wished she had Susan beside her, to brainstorm what to do next. The policewoman knew so much more about the corporate world than Carolyn. Why would a company in chaos *move* its records? Perhaps to destroy them, as the pessimistic young man thought. Perhaps to look through them, for whatever it was the Vandals, Not-Vandals, and Carolyn herself were after? The second possibility left them extant, so she would use that as a working hypothesis. If you were a company sorting through decades of records, where would you do it?

Perhaps not in your main headquarters. She rested her face in her hands. There were the three other sites, but she had no clue where to even start looking in them. She stood. She was in this building now; she should take advantage of that. She could discount the basement given that had they been somewhere else down here, surely their location would not be so unknown.

The opposite out-of-the-way place would be up. She leaned her head back and regarded the staircase above. A small rectangular hole vanished upwards, around which the stairs circled. She had not paid attention to how many floors there were; it seemed to rise forever. She hesitated, thinking perhaps to find a lift, but she was currently undetected in the stairwell. Who knew what unexpected scene a lift door could reveal. She started climbing.

* * *

She had lost count of the number of stories she had climbed. The interior of the stairwell had no clues as to what floor she was on. The staircase no longer rose endlessly; she could almost imagine she saw a celling above. The very, very top might not be it. She could start checking to see if the decor changed at any point.

She cracked the door open and peered through. She looked into a clinically white and austere corridor with an eyewash station visible off the left-hand side. Labs. She closed the door and climbed again. The next three floors were nearly identical, save for slight shifting of the eyewash station to handle differently placed doorways.

She stared up the slight rectangular hole where the stairs folded back on themselves. She was nearly at the top; it could not be more than three or four more stories to the now clearly-visible ceiling. The next door revealed a red-carpeted corridor with wood-panelled doors. The air was distinctly perfumed, floral with a hint of a cinnamon-like spice, a contrast to the stark cleanliness of the labs below. Voices came from somewhere in the corridor. She pulled the door closed and raced up to the next floor. Maybe they had not seen.

This floor had a blue carpet plus the same wood-panelled doors and floral scent. There was just one more floor; she might as well check. The top floor seemed unfinished—a rough cement floor and dingy walls. She slid through the door. The corridor appeared abandoned. The door clicked shut. Adrenaline shot through her—she should have checked to make sure the door would open again! She tried the handle, and it opened. She breathed a sigh of relief, then moved cautiously down the corridor. There were no doors until she reached the very end, where a single metal door opened to the left.

It had a long push-bar across the centre, like often seen on emergency exists. She pressed it in, half expecting an alarm to go off. None did. At least where she could hear.

The door opened to a brightly lit room that appeared to run the whole length of the corridor. It was empty, save for one man on the far side. He turned as the door opened. She backed up and let the door close. Uncertainty shivered her legs. Had he seen her? Footsteps sounded, running towards the door. Adrenaline spiked again, and she ran down the corridor, towards the stairwell.

She reached the door and plunged through, then ran down the stairs. Her breath came fast. What to do? She was a sitting duck in the stairwell. She opened the door onto the blue-carpeted hallway and turned randomly left. As she walked she berated herself; she should have gone down to the labs. She understood that sort of location. These two floors were some kind of corporate offices. She would be out of place.

A door with a small skirted shape—toilets! She entered the ladies' room and retreated to a stall. She leaned against the wall, panting. Fear still shivered her limbs. She was not made for this sort of thing.

Calm down, she told herself. Someone had seen her, but they could not know who she was. She was just a person out of place, that was all. Her desire to catastrophize failed; she could not even imagine what it was she feared. The Vandals had already had her fake-die and destroyed her life. There was nothing else to fear. Unless they somehow got hold of Ellen—for what, Carolyn could not imagine. But she could not imagine everything that had happened so far, either.

Resolve filled her. She remembered her anger in Uncle Keith's conservatory, and her confidence. She would sort this, whatever was happening. She left the toilet stall and automatically washed her hands. As she dried them, she formulated a plan. She would be bold, just as she had in the basement.

*   *   *

Now that she was looking to run into someone, there was no one to be found. She circled the whole blue floor without encountering anyone. She went again, trying all the doors and finding them locked. There had been people on the red floor. She returned to the staircase and descended one story.

This corridor was empty, too, but at the far end of the building she finally passed someone. A young woman in a business suit held a stack of papers poorly balanced against her chest. They started to slide. Carolyn raced forward to catch them.

"Thanks!" said the woman. She put the pile on the floor. "Tried to take too much." She put her hand on her hips and stared at the pile.

"Can I help?" Carolyn asked. Not quite her plan, but getting on someone's good side could not hurt.

The woman bent down and lifted half the pile. "Sure."

Carolyn lifted the other half and followed the woman into a small window-less office. She placed her pile beside the woman's on a desk.

"Thanks," said the woman again.

"No problem. Um." *Speak!* Carolyn told herself. "Could you direct me to records?"

The woman's eyes narrowed. "Why?"

"I'm an intern." That had worked in the basement. They must actually have interns. "My boss sent me up here to help."

"You're on the wrong floor. You need to go up one more."

"But there's no one there!" The frustrated sentence popped out before she could stop it. A half second later, she realised it fit. An intern sent to 'help' with records could easily be annoyed to find the empty corridors she had.

"Did you knock?"

"Um, no, where?"

"Any of the central doors. Go back up and knock."

"Okay, thanks." Carolyn backed out of the room, leaving the woman lifting the top paper off one pile and frowning. She headed boldly to the lift this time. An intern planning on knocking on a door would not be lurking about in a stairwell.

The lift doors opened. She stepped out into in the empty blue corridor again. A door stood directly across from the lifts. She took a breath and knocked.

Chapter Seventeen

Nothing happened. She bit her lip, wondering if she should go back down. But just before she turned to press the lift call button, the door opened.

"Yes?" said a young man. He was plastered in sweat. He wore a suit, but with his tie nearly undone and his shirt open halfway down.

"Um, here to help?" She had not meant it to be a question.

"Thank goodness!" He opened the door wider.

She stepped into a room as large as the one on the unfinished floor above, but carpeted and filled with row upon row of thin metal folding tables. On each table was a central rim of papers and what looked like lab books. One to three people per table stood grabbing from the pile and flipping through. When finished, they dropped whatever they held and kicked it under the table.

"You can start over there." He gestured to a table where the central pile was still neat and there was no heap of discards underneath. "Do you know what you're looking for?"

"No?" Her voice rose in a question again.

"Anything by the Schwarzes. If you find it, bring it there." He gestured to the far end of the room. Carolyn looked over the expanse and could just make out a table with three grey-haired, suited men each slowly reading a lab book.

Her heart jumped into her throat. They must be looking through her parents' lab books! What was inside? She itched to go over and see. Her mind turned to her mother's ID, resting conspicuously on her chest. She forced herself to not look down. Was it lying picture-side out, or the back, with all the writing? Either would be awkward.

The young man gestured again to the untouched table. "I'd recommend removing that coat. What heat they've on rises. You'll find it much warmer here."

She moved off quickly before he could inspect her attire any further. She positioned herself at the centre of the table and reached for a lab book. So, Vivcor was searching its own records. Had they seen the ones she had, from Paris? Why all these tables—did their records have no organisation?

She opened the lab book. She knew immediately it was not one of her parents', as the handwriting was different than that she had grown used to on the train ride from Paris. She could distinguish her mother from her father now. But a random intern would not be able to make the determination so quickly. She flipped through the book, looking for something identifying. Neither the front nor back covers had a helpful name. Poor lab practice. One should always name and date lab books.

The young man who had greeted her appeared at her elbow. "That's not one."

"How can you tell?"

He took the book from her and flipped to the back. "Schwarz lab books will say somewhere—in the front or back covers." He scanned the back cover, then the front and back of the stiff page at the very end, and repeated this in the front. "Something with their names or even just S-lab."

"Okay, thanks." She watched as he dropped and kicked the book. "Are we really just leaving them there? Isn't that a bit … um …" She trailed off. It was not her problem if Vivcor wanted to leave their scientific records in dozens of untidy piles, was it? Yet if she had a soft spot for any company, it would be the one who had helped her change her name and erase her past. Never mind it also benefited them for Baby C to be forgotten; they had helped her, and seeing all these people treating their records like this distressed her.

"Not our problem. If we find the Schwarz stuff, they'll be able to hire an army to deal with the mess; if not …" He shrugged, similarly fatalistic to the overalled man in the basement. "Let's hope we find it."

He left her again, and she reached for another book. Five books later and she found one with *S-lab* scrawled on the first, stiff page. Her breath caught. She kept flipping, looking at the notes. Neither of her parents' handwriting, so it must be a lab member. They were building artificial chromosomes in mammalian cell lines. She checked the date; four years before her birth. It would be preliminary work leading up to the big hoax. Perhaps they even thought it would work at that point. When did they decide to fool everyone? How many people were in on it?

She had never considered that before. More than her two parents were involved. Had they hid their deception from their lab members as well? She put the book down. It was not the time frame she was interested in, nor likely to be what the people at the end table were, either. But a random intern would not know that. She lifted the book and walked across the room.

The stacks of lab books at the end table was smaller, but still impressive. This would be her parents' stuff. She *so much* wanted to look herself. At least these old men placed their inspected books in neat piles, if still on the floor. Perhaps she could come back when the room was empty? Although if everyone was looking for the same thing, they would be unlikely to discard something relevant. Yet potentially she had knowledge from those scanned books they didn't; she knew she needed to see what had happened just before her parents left Bucharest.

She held out her book, and one of the men nodded wordlessly at a pile. She placed it down and started back to her table. Her back tickled. That had been silly, she realised belatedly—she still wore her mother's ID. These men would be higher up in the organisation and more likely to care who was around. But they had not noticed anything, and she had gotten a look at the table and the books. She reached her table and looked through more books.

Hours went by. She felt trapped in this stuffy room with her clothes clinging to her sweaty skin. She knew where she needed to look, later. But she could not explore in this room full of people. She also could not leave without being conspicuous. Sandwiches appeared around midday, rolled in and left on tiered carts in the approximate centre of the room. She ensconced herself at the edge of the large crowd of mostly young searchers as they sat on the floor eating. She listened to the conversation around her, occasionally smiling, but no one seemed to mind that she did not join in. Most people were sharing stories of job searches, helpfully passing around business cards of potential contacts.

Vivcor did seem on its last legs. Could they really hope to find something in her parents' research that could save them? Why *now*?

Her questions remained unanswered as two middle-aged women in suits shooed everyone back to the tables. She spent several more hours—after a short comfort break—clearing about half the books from her table with no further finds from the Schwarz lab.

"Thanks all!" shouted one of the suited women from after lunch. "Tomorrow morning, sharp, and we may actually finish this week!"

General rumbling answered her, and people moved towards the doors. Carolyn joined the exodus and stopped in the toilets like several others, except she stayed in her stall until everyone else had left. She leaned her head against the cold metal wall. Back to the room, if they had not left it locked.

She waited what seemed another fifteen minutes or so, just to be sure. Her sweat had chilled on her skin by the time she crept into the hallway. The lights were off and burst to light in response to her motion. She jumped, but the corridor appeared empty. She tried the first door. It was locked as they had been this morning. All the remaining ones were locked as well.

She blinked back unexpected tears. So close!

Determination filled her. There had to be another way. She climbed back up the stairs to the unfinished floor, ideas half-formed. Maybe she could somehow see downwards, or otherwise figure out from that floor how to get past the doors on the one below.

Lights in the upper corridor were on as before, and the large room still brightly lit. She pushed the door open further.

A man stepped from behind the door. "Becca?" he said, in a breathy, shocked voice.

Chapter Eighteen

It was the side-burned man. She took a step back.

"My God, you must be Carolyn! You gave me quite a fright. What are you doing *here*?" The breathy fear in the voice immediately triggered memory of that phone message: it was the same man.

"Who are you? What are *you* doing here? Where are your friends?" She stood tense, ready to run. Was this the same person who had been in this room before?

"They're no friends of mine! They …" He looked back and forth. "You had best come in." He stepped further into the room. "This used to be some kind of testing facility, I think. There appears to be no surveillance."

So this was the Informant. She stepped fully inside. "Did you write that note? Who *are* you?"

He closed the door and sunk to the ground against it. "I was Becca and Manny's friend. I *knew* it was no accident. I *knew*."

Becca and Manny: Rebecca and Manfred Schwarz, her parents. He had been in the photo of them. "You were a student with my parents?"

He nodded. "I went into academics; they went corporate." He laughed, hollow, strained. "They had so much more money than me. I remember that first year out of grad school, and they got a bonus each—enough to buy a full entertainment system. A giant TV and stereo and everything. It seemed so extravagant. At the time I was barely making ends meet as a postdoc."

These reminisces were fascinating, but a bit beside the point right now. "Did you send me those messages? Did you photograph all those lab books?"

"You found that!" His face lit up. "That's when *they* arrived. I shoved the elly-book somewhere ..." He screwed up his face, as if trying to remember.

"Behind a fridge," she supplied, impatient. He was garrulous but his words seemed content-free. "*Who* arrived? What is going on?" Her fists were clenched. She breathed deep. This person had to know the most of anyone she had yet encountered. Just because he was not spilling it all as fast as she wanted—when they had only just met—was not a reason to get so frustrated. "Wait, what wasn't an accident?" He looked so small, there on the floor. She felt as if she were towering over him. She sat down as well.

He closed his eyes. His face was tense, jaw tight. "Your parents' death."

Her world shifted. She placed her palms on the floor beside her, as if she could hold on. "My parents died in a car accident on holiday."

"That's what everyone's supposed to think. But it wasn't a holiday—they told me they were running." He pushed his hands against his closed eyes. "I think they hoped it would blow over, and they could come back like nothing had happened." He lowered his hands, revealing eyes brimming with unshed tears. "Instead they didn't come back at all."

"Stop talking around the edges and tell me what's going on!" Her shout echoed around the room. Her cheeks heated; she hoped this 'testing facility' was surveillance-free as claimed, plus reasonably sound-isolated from the rest of the building.

"Sorry," he mumbled. "It's just I barely know myself." She fought back the urge to snap at him again, biting her lip instead. He twined his hands in each other. "They found something they were scared of. Before you were born. Before you were *made*. Or supposedly made, I guess. They never said what, just that it changed everything and they didn't think the corporate world could handle it. I think ..." His constantly moving hands stopped, and his knuckles whitened. "I think they were hoping to pass it on to me, later, and let it come out in academia. They told me—'suggested'—areas I might want to take my research. I think they were laying the groundwork. But already there was the creep of the corporations, and the Universities were becoming less and less independent. They waited too long."

He brought his taut hands up to his chin. "They were going to move again. Someone had figured out what they were doing in Paris was just distraction. They were going to holiday out their final days of their Paris contract, then move on to Lihue. I don't know why they thought distance would help. It was still the same company. But I think they were worried—before they left they told me to look to you for the answer. That they'd passed it on to you." He turned hopeful, red-rimmed eyes to her.

She shook her head sadly. "I don't have anything."

"Are you sure? They must have left something—why did you end up studying mitochondria?" Why hadn't she studied *English* like she'd originally planned? Her life could have been so much simpler. "That's where they were pushing me. That and something about the chromosomes they were building for you." He grimaced. "Supposedly building."

"I had no idea my parents had anything to do with mitochondria. I picked that to be as far as possible from them, when I found myself a molecular biologist. I intended to avoid biology, but …" How had she ended up there, anyway? They had required first-year students to take courses outside of their subject, and that biology class was so interesting. It had not even mentioned her—the Hoax was a thing of the past, especially when a few years later the first actual synthesized humans were born. The fascination from back when she thought her parents had been true pioneers rekindled. The fact that her parents' deception had been overtaken by the progress of science made it seem like it would not be so bad to pursue biology after all. That vow by a stunned sixteen-year-old seemed teenage folly. She wasn't going to let her dead parents' scientific misdemeanours define her whole life.

"You mean you *don't* have it?" His voice was shocked. "But why … why are you here?"

Her parents' scientific misdemeanours were *destroying* her life. "They ransacked my lab, my flat, and then without ever talking to me they manufactured a fake death for me." Her voice dripped with anger. "They gave me no choice. Who *are* these people?" She stared at him; if she could have bored the answer out of him with her eyes, she would have.

"I thought they were Vivcor. But I escaped when they got chased away breaking into here. So I don't know."

Her heart dropped. He didn't know. But he knew far more than she did. "How long have you been here?" She tried to review the timeline in her mind. What day was it now, Tuesday? "They followed me to Stirling on Friday—"

"No, they followed you to *Paris* on Friday." He sat up straighter, as if the correction had reinvigorated him. "But they couldn't find you and came to Bucharest, hoping … I have no idea." He slouched again. "But I guess they were right, as you showed up."

She sucked her lip, distracted by the puzzle. "No, they did follow me to Stirling. People came to my Uncle's place, asking after my parents' stuff—looking for prox cards." At the thought, she touched her mother's ID hanging on her chest. "These, I guess. I thought maybe they wanted them to get into Vivcor Paris, like I did."

"They didn't need prox cards. They got in on their own, when they found me."

She stared at him. "How many people are after whatever-this-is?"

"What do you mean?"

"There was another group in Paris. I thought it wasn't your people, as they'd gotten into the Paris lab. But maybe …"

"No, my people never went back to the lab in Paris. They looked through everything there and took what they wanted when they found me. That's what had them heading to *your* lab."

"Not that you told them I had this secret?"

"I didn't!" He straightened again in his self-justification. "I didn't breathe a word of that."

"What about that message you left?"

"That was to you—I was trying to get you to give it to them, so they'd leave you be. They went after you because of the mitochondria—I think they figured out that was the direction from the lab books; I didn't get a chance to read them. You're famous in that area."

Famous? She would not have said that, although she had given a number of keynote talks and was reasonably well-known. "Okay, I think we need to lay things out here." Her hands itched for something to scribble on. She made do with making gestures on the floor. "The people I'm calling the Vandals broke into my lab, looking for research?" He nodded. She patted the floor to her right. "Some other people followed me to … no, showed up in Stirling. They might not have known I was there. They were looking for prox cards." She reviewed the event in her mind; they had *not* asked after her at all, just her parents' things. She patted directly in front of her. "Another group grabbed me and Susan in Paris. They were looking for … me?" They had used some kind of genetic test on Susan, the cylinders for which Susan still had. She patted to her left. "You're saying the Vandals weren't either of these two." She put one hand left and one centre. She frowned. "Could these be the same people? But the Stirling people were looking for prox cards, and the French had access to the building."

He was staring at her chest. "Access isn't all the prox cards might have. What is all that?" She lifted the card. It had been lying with the written side out. He leaned forward, squinting. "If anything might be considered a secret code, that would be it."

She took a closer look at the writing on the card. She had dismissed it almost immediately as meaningless without other information—whatever everyone was searching for. This couldn't possibly be it, could it? That she'd been wearing for days? It had a pattern, with small sets of numbers clumped together followed by long strings of letters. Every string of letters started with an 'M'. She stared further at the letters, an idea growing in her mind. There were letters missing: no O's or U's, when you might expect vowels. The idea solidified. She looked carefully: no X; no B, J, or Z, either. "In what language does every word start with M?" she said to herself, but the side-burned man was a biologist and it would have meaning to him, too.

"Proteins?" He reached for the card.

It wasn't precisely true; only about forty percent of proteins retained their initial methionine, but because the signal in the DNA for the start of a protein also coded for that amino acid, abbreviated 'M', all proteins began life starting with M. And the bit of DNA known as the *coding sequence*, if translated directly into the protein it referred to, would start with M. "Translated coding sequences, at least," she clarified.

He hummed agreement. "What are these numbers? Protein designations? Chromosomal location?"

"If they're proteins, we can BLAST them and find out." She referred to the bioinformatics program that searched the world database of sequences. It would let you know what any protein or DNA sequence matched.

"From *here*?"

"Well, I didn't mean that." But the elly-book's text interface would be perfect for BLAST. She even knew that web address by heart. "I need to get back to Susan and let her know what I've found. What are you, um, doing?" This fellow was on her side, right? Not a plant by these non-Vivcor-Vandals?

"I'm just hiding. Who's Susan?"

"A friend." She realised she did not even know his name. After the initial rush of information, she was getting wary. "How do I know I can trust you? What's your name?"

"My name?" He blinked, as if surprised she lacked this basic information. "I'm Ethan. Ethan Boltzer."

"*Professor* Ethan Boltzer? Of the Boltzer Process?" The Boltzer process was the standard method of transforming—inserting genetic material into—mitochondria within eukaryotic cells.

"Yes. I told you Becca and Manny suggested I go into mitochondria."

Professor *Boltzer* thought she was famous in the field of mitochondria? He was arguably the most famous person in the field himself, even if she only knew him by name. She blinked, assimilating this information. But if they were in the same field, and he knew her parents … "Why didn't you ever get in touch?"

"I didn't know who you were! You vanished once the Hoax came out."

She supposed she had. Vivcor had helped. "Are you sure your people aren't Vivcor? Who else would know who I am?"

"I have no idea." He spread his hands.

They were not getting anywhere with this. They needed to go through events more systematically, and probably not on the top level of Vivcor's Bucharest Headquarters. "We need to get somewhere we can talk better." Although they were so close to her parents' lab books, stacked neatly on a table one floor below. "But I'm in the building now. I came here trying to find the research that preceded Paris—that's what your friends wanted—"

"Not *friends*!"

"Your—captors?—wanted too. But Vivcor is close; they think they'll get through the books this week." She stood, looking around, as if she could get downstairs *right now*.

Boltzer levered himself up as well. "How do you know this?"

"I was pulling out lab books with the rest of them. They think I'm an intern; they have tons of them at the job. It's just below here." She stared at the floor.

"Just below?" He started moving, walking faster until he took up a jog.

She chased after. "What are you doing?"

"I've been stealing food from the kitchen down there." He pointed at the floor. "I knew they were prepping for a crowd; I didn't realise what was going on."

"You can get into the room below?"

"Maybe." They stopped at the far end of the room, approximately where she had seen him when she first looked in this floor. He pushed open a small door that lead to a narrow room with a hatch at the end. "It's some sort of dumbwaiter, like a very short lift. Probably not meant to hold people, but it worked." He squatted down and opened the hatch. He frowned. "Maybe one at a time." He shuffled in, pressed a button on the wall, and closed the door.

She stared, feeling queasy. Climbing down the outside of a building with Susan did not bother her, but stepping into some sort of miniature freight lift clearly not made for people was distinctly scary. But everything about her life this last … not even a week yet! … was scary. She was not sure how to tell when the lift was back. She tried the door. It opened. She crawled in and folded over, feeling squished as the ceiling pressed hard against the back of her head. She pushed the same button as had Boltzer. She closed the door, and the lift immediately dropped, fast. Panic spread through her. She knew she shouldn't have gotten in! She placed her hands on the floor and squeezed her eyes shut.

It stopped with a jerk, and the door popped open. The panic washed away, leaving her feeling tired and embarrassed, even though Boltzer could not have known her reaction. She crawled out and stood up. They were in very narrow kitchen, mirroring the narrow room above. Narrow closets nestled next to the dumbwaiter, with the remainder of the room lined in shiny black counters, burnished steel hobs, and racks of pans hanging from the ceiling. It smelled of garlic and bacon; she surged with brief envy for whoever had gotten to eat the cooked food, rather than the dry sandwiches fed to the searchers. Boltzer walked in front of her towards the door to the large room. A small window sat high in the door. Through it she could see the end table, with one of the elderly suited men sitting in a small pool of light.

She put her hand on Boltzer's shoulder to halt him. "Someone's there," she whispered. Her mind screamed in silent frustration. So close, and yet there was someone in the way. She had an urge to step out there, knock the old fellow on the head with one of the pans beside her, and run off with the lab books. It seemed to be the way everyone was acting around this.

As if the universe could read her mind, two figures appeared behind the old man and grabbed him. He struggled briefly before going limp. The woman holding him dropped him to the floor, then joined the other one—a man—at the table. Together they swept the lab books into a sack and vanished into the darkness.

Chapter Nineteen

"No!" She could not keep this scream silent, though she brought it down to a whisper. The old man lay still. Had they killed him? Fear spiked through her, chased immediately by guilt that she had thought of attacking him as well. But she would not have. Not really.

Boltzer stood on his tiptoes and looked through the window. "What happened?"

Unthinkingly, she pushed open the door from behind Boltzer and crossed to the old man. She squatted. What were you supposed to do, check for a pulse? Scenes from crime vids passed through her mind. She pressed on his neck, but it was just warm flesh; she had no idea how to tell. He moaned and shifted. Well, that was life.

Immediate terror that she had just seen someone killed faded, leaving delayed, more practical concerns. Surely this room had surveillance. She scanned the walls and ceiling. Too late now. The table, so recently full of her parents' lab books, was empty. Her heart fell. She slumped fully to her seat on the floor. Someone—who clearly was not Vivcor—now had what they were looking for.

"They didn't take everything," Boltzer said softly, clearly thinking along the same lines as she. He was on his hands and knees under the table, where there stood stacks of discarded books.

"Those will be ones that don't matter," she said.

He picked one up and opened it, then another, and another. "No, they *will* matter. This is the exact time frame we need." He passed her the book. "Look at the date."

It was two years before her birth, which would be a good five years before the move to Paris. She wanted to see what they found right before the move. "But that's ..." She trailed off, remembering Boltzer's words. Her parents had found something *before* she was born ... or designed ... or supposedly designed. Vivcor didn't know that, nor might anyone else who looked at the Paris books. They would come to the same conclusion she had, that the secret lay right before the move to Paris. Hope rose. They had an edge no one else did. "Let's take these and get out of here."

There were not that many discards, but too many to just carry. The kitchen would surely have some sort of bags. She stood. "I'll look for something to put those in." She went back into the kitchen.

That giant pot would fit some books, but was not practical. She recognised an edge of humorous hysteria to her thoughts, like when she had found her parents studied mitochondria. She took deep breaths. She opened one of the narrow closets beside the dumbwaiter. Various sized sacks poked from pockets inside the door, mostly handled ones like used by grocery stores. She grabbed a handful and jogged back to Boltzer.

They quickly filled four sacks with lab books. She stood, heavy books hanging in a sack off each arm. "How do we get out?"

"I don't know." He lifted two sacks of his own. "That's why I stayed here. I'm scared *they*'ll be watching for me."

"We need to be somewhere else. Come on." She led the way to the kitchen and the dumbwaiter. She went first this time, swallowing her fear.

She dragged the sacks of books out, then finally helped Boltzer stand. "We can talk more freely here," she said. "We need to move. They probably don't have a live feed, otherwise someone would have come in long before we left. But they're sure to view their vid in the morning."

"But what do we do?"

She hefted two sacks of lab books. "We walk out. Then deal with whatever happens." The plans she and Susan had viewed showed several exits other than the main one. There was one accessible from near where records *should* have been. "Follow me."

*   *   *

Getting Boltzer down the staircase took longer than she expected. For a man who could kneel and squeeze himself into that dumbwaiter, his knees did not take stairs particularly well. But she did not want to risk the lifts. They wound their way through the basement level. Movement-triggered lights surrounded them, making her nervous, but they could do little about it. The two attackers—Vandals, maybe?—had already insured that Vivcor would be scouring their surveillance feeds.

She found the exit still extant, halfway up another staircase on the far side of the building. She stepped out of the door into the cool night. They were in a narrow alley between Vivcor and the shops on the other side of the street. The air smelled strongly of fresh bread. It must be late enough that the bakeries were already up and cooking.

Her mind spun. They would be tracked by any vid around. She could not head straight to the anticorporate headquarters. Their hosts had provided them with a list of surveillance-free routes and autocab companies. She had not imagined needing to use them on this trip, planned to be more reconnaissance than anything else, but the information popped up as if burned in her brain. "This way," she said, and Boltzer followed her out of the alley.

The streets were lightly trafficked. She saw no autocabs, so kept walking, heading past University libraries and museums into the anachronistic centre of town. She descended into the Metro and absorbed in an instant the map that she had poured over on the way out. She handed her ticket to Boltzer. "Go through, I'll come on the same ticket." They were already on the run, jumping the Metro gates could hardly make things worse.

He obediently took the lead, but hung back to wait for her to pass once they were through the gate. She had a fleeting thought that he was following her in the same trusting-puppy manner that she had followed Susan in the train station in London. And like Susan, she did not speak save for a few quick words of explanation when she was sure they were surveillance-free. Several Metro, cab, and bus rides later, she was more confident they could vanish into the increasing morning commute. She took a safe autocab out to the same hill she and Susan had climbed the day before.

They topped the hill. "What in the world?" Boltzer said.

"My reaction as well." She kneeled and lowered herself over the edge. She reached up towards him. Given his trouble with stairs, this drop and slope may not be the easiest journey. But with an occasional steadying hand from her, they made it to the grassy bottom safely.

She led the way along the trail heading into the centre of the marsh. Rustling sounded ahead of them, then six figures appeared from the brush. She recognised the teen who had led them in last time, as well as the older woman with tightly curled grey hair who had oriented them to the hideaway and the town's safe routes.

"Who's he?" said a tall, scar-faced man standing next to the grey-haired woman.

"A friend," said Carolyn, haltingly. Her earlier confidence as she led the way through the city dissipated. She wanted Susan beside her to deal with this underworld.

"No friends have been mentioned." He stepped forward threateningly.

Chapter Twenty

"I just found him," Carolyn said. "He was kidnapped by the people chasing us." When already snooping into her parents' research. His story of being caught in the Paris lab replayed in her mind. He had not explained why he was there. She stared at him curiously. Suspiciously.

Boltzer put down the bags he was carrying. "Look, I never wanted any part of this. All I want to do is get home." He spread his hands. "I can just leave …" He trailed off, his tone of voice suggesting he had no idea how to actually do that.

"Not so easily you can't," said the scar-faced man. "Is this fellow of use to you?" He directed the question to Carolyn.

"Yes." Boltzer's knowledge of her parents' history had already given them key insights. He had clearly not shared them with anyone else, since Vivcor and the Vandals both targeted the wrong time-frame.

Susan appeared, jogging down the path. "Where have you been?" Her voice was a near-shout, tinged with the anger of panic just relieved. She paused. "Who's that?"

"Professor Ethan Boltzer," Carolyn answered. "My parents' old friend; the same one who left me those messages."

"He's going to help you?" scar-face asked.

"Yes," Carolyn answered confidently. Boltzer looked scared, and sympathy for him rose. But first she wanted to hear more of his story.

Susan stepped closer. "The man in the video! You were with the Vandals. Who are they?"

"I don't know," said Boltzer. "Apparently *not* Vivcor."

Susan frowned. "Well, that's something."

The scar-faced man stood with his arms crossed. "I'm not happy about this, but he's here." He squinted. "Follow me." He led them off a side path, to a small hill near a pond, where some dirty grey blankets sat rumpled underneath a tree. "You can stay here. I expect you all out tomorrow."

"Our stuff—" Susan started.

"*You* can come get it." He spun on his heels and left.

Susan turned to follow them. "Back in a jiffy."

Carolyn stared at the dirty blankets. Apparently she and Susan rated better digs than the pair of them and Boltzer. She sunk to the ground, her sleepless night catching up to her. She opened her eyes, suddenly realising she had drifted off. Boltzer lay beside a tree, curled around a blanket, snoring lightly. She closed her eyes again.

* * *

The next thing she knew, the light was fading, and she was getting chilly.

Boltzer shifted beside her, obviously also just waking up. Susan leaned against the tree, still dozing.

Boltzer stretched. "What *is* this place?"

"Anticorporate hideaway. Or off-grid group, or something. Susan says there are communities all around the world. I had no idea."

"Me either." Boltzer pushed himself more vertical. He pulled a lab book from one of his bags. "Sorry I drifted off. I wanted to take a look at these things." He opened it and squinted. "We can't read them in this light. But we might be able to tell the dates. We can at least sort them."

Carolyn scooted over. She itched to know what was in the books, but he was right. With concentration, she could just about make out the date line starting the entry. "What dates are we looking for?"

"About two years before your birth, and before. After that they'd already started the distractions."

She pulled a book from the bag nearest her and found a date. "So, you know my story, mostly. I was minding my own business when your friends—" She held up her hand. "Sorry, captors—broke into my lab and started this whole business. I went to Stirling, at which point they 'found' my dead body back in London. I went to Paris, then we met here. What about you? What were you doing in Paris?"

"I'm retired." It sounded almost like a defence. She just kept checking lab books and stared at him. "So, I was minding *my* own business, when out of the blue I get these government agents at my door. They were investigating your parents' death—it happened in the US, you recall—a cold case." He frowned. "I don't properly remember the details, but new information had come out of Vivcor's reorganisation, so it was open again. The agents thought someone might have been after them for their Final Findings."

"Government agents knew about that?"

"They knew far more than I'd expect about our field."

Ethan Boltzer had just grouped his research with Carolyn's, under the term 'our field'. Her chest fluttered in pride. Although would her research career ever restart, after her 'death'? She set her jaw. She was going to get her life back. "How did you end up in Paris?"

He buried his face in his hands. "I was an idiot."

"Yes?"

He lifted his head. "This fellow—he *said* he was a reporter. He said the government would cover it up, and he wanted the real story of the Schwarzes to come out. I should have gone to the Feds. But he made me suspicious of them ..." She raised her eyebrows, and he sighed. "He said he had gotten into their last lab in Paris and wanted my help understanding what they were doing. At least I had the sense to not point out the time frame. Plus, I thought they might have brought their old lab books with them. But anyway, I snuck away with him to Paris and helped him photograph all their lab books. Or almost all. Then Vivcor showed up … wait, not Vivcor. What are you calling them?"

"The ones you were with? The Vandals," she said.

"Then the Vandals showed up. I was in the lab taking photos with the elly-book and hid in that hole beside the fridge."

Exactly where she had hidden. Behind him, Susan stretched, then came to squat quietly beside them.

"At the time I didn't realise all the books were too late, so I hid the elly-book, where I guess you found it." He gave a crooked frown, as if contemplating on how their paths had wound about each other. "They found me, but not the elly-book. Well, then you know what happened."

"Not really," Carolyn said.

"They dragged me along with them to your lab in London, wanting me to identify anything that looked like your parents' work."

"*Was* there anything?" She was honestly curious. She had been convinced her work was nothing like her parents'. But that was before she found out they had been studying mitochondria.

"No. That's why I left those messages. I figured you were hiding it …" He trailed off, his voice suggesting he still held some hope that she had held back.

"I really do know nothing," she said. "What happened next?"

"Then they went to your flat. Didn't find anything there. So, do you have a roommate or something? An artist? They were wondering about that."

"I have a *daughter*," Carolyn said. Stunned surprise—how could he be ignorant of such a basic facet of her life—was replaced quickly by relief. Boltzer didn't know her; he clearly wouldn't know she had a child. But that meant the Vandals also did not know about Ellen!

She had not realised how much she was worried they might still try to harm her daughter until the worry was relieved. Then panic followed—she had just told Boltzer. Was that wise? Yet the more he spoke, the more she was convinced his story held up. "But surely they would have learned that, when they fake killed me. The effort that took …" There were newspaper articles with fake quotes and everything. Not one of them had mentioned Ellen. She had not focussed on that, other than distracted relief, instead reeling from the thought that everyone believed her dead.

Susan shifted. "Maybe some turncoat in the force had an ounce of ethics left."

"What do you mean?" Carolyn asked.

"If the Vandals knew about Ellen, they might have tried to take her hostage or something. Sounds like someone tried to keep her hidden. She lives with you full time?"

"Fifty-fifty shared," Carolyn said. "But I'm her address for school."

"So it would be in records, easily. You're right. It needed to be hidden." Susan frowned.

Carolyn's chest tightened at Susan's worried look. "Does that mean Ellen is in danger? What if they find out? We didn't know she was supposed to be a secret!" She should have just gone off with Ellen to one of Bae's anticorporate hideaways, instead of haring off on this chase for who-knows-what. Now she was a continent away and did not know what was happening.

Susan put a hand on her elbow. "Keith and Bae will make sure she's safe." Carolyn nodded, but the words felt cold comfort. She wanted to be home with her daughter. She wanted her life back. She wanted … she wanted none of this to have ever happened, and that she could never have.

Chapter Twenty-One

Carolyn woke in the morning, stiff, but probably less so than Boltzer. He followed her and Susan to the facilities with a limp that had not been there yesterday. Susan went to fetch breakfast from the central mess they were still banned from, and Carolyn and Boltzer returned to their tree and the lab books.

The sky was grey, but the air still. A continued cacophony of birdsong suggested the birds considered the overcast a very long dawn. They had finished sorting the books last night. They sat and, with silent agreement, started reading through the relevant ones. There were six different people working in the group at the time, her parents and four other sets of handwriting. Her parents were involved in all the projects, but in such small snippets it was nearly impossible to follow any one strand. Of the other four, two were synthesising chromosomes, one was concentrating on a problem with the telomeres of the synthesised chromosomes, and one was doing an apparently wholly unrelated project on mitochondria.

She barely noticed when Susan returned, other than to absently eat the provided breakfast. She licked her greasy fingers, realising she barely remembered what she had eaten. It had been some sort of fried potato strips, perhaps with cheese?

"What do you suppose this means?" Boltzer asked, pointing out a notation in one her father's lab books: *See E-B.*

"I don't know. I've found them too. But only in my parents' books." She frowned. "It's not you, is it? That's your initials."

"No …" He stretched out the word as if unsure. Then he shook his head decisively. "This is before they started hinting to me. It *can't* be."

She turned the page in the book she was looking at. It was hard, trying to find some kind of overall pattern from the minutiae of daily laboratory records. The current writer—the mitochondria researcher—had just spent two weeks failing to get a PCR reaction to work. PCR, or Polymerase Chain Reaction, was a method of copying small bits of DNA. Pages and pages of identical gel photos, blank save for the ladder—DNA fragments of known sizes that you could compare your DNA bit to—gave her little info.

Then: *PCR worked!!! B&M taking over. Gels in E-B.* "Hey, I've got the same notation here."

Boltzer leaned over to look. "Hmm. 'B&M taking over' must mean Becca and Manny—your parents. What's the date? Maybe we can cross reference."

She told him, and he reached back to the pile of her parents' books. He flipped through. "No, nothing about that here at all. Manny's just dipping into the chromosome stuff then." He picked up another book. "Becca too." He frowned.

"Where is it then?"

"E-B?"

"But where *is* that?" She turned the page in her book. There was a two-month discontinuity between the last entry and the next. "We're missing stuff."

"Viv—the Vandals did take some books."

She tapped the space above the current entry, where the missing two months should be. "No, we're missing stuff right here, in these. No one would just randomly write in a different lab book. E-B must be another *kind* of location." She closed her eyes. "Docfilm didn't exist yet, did it?"

"No," he said in a drawl, as if thinking as he spoke.

She remembered her hope in the Paris lab of finding an e-lab notebook. She snapped her eyes open. "But elly-books did! E-B! There *is* an e-lab notebook somewhere." She frowned. "Vivcor can't have it. They would not be searching through reams of physical books if they did."

"That's the type of thing they *would* have taken to Paris," he said. "Elly-books aren't cheap now. Back then it would have been a pretty expensive piece of kit. You wouldn't just leave it in some company storage somewhere."

"Did you find one in Paris?"

"No, and my 'journalist' didn't seem to have either—I can't imagine why we'd be photographing the physical books if he had."

"Why are you sure he's *not* a journalist?" Susan asked. Carolyn silently seconded the question. Boltzer had not explained that.

He looked confused, momentarily. "No news story has come out, has it? He said he was going to blow the whole thing wide open. It's been weeks. He could have gone back for my elly-book, but didn't—you found it."

Carolyn said, "Weeks? I think we need to go over the timeline in more—"

"What happened to *him* when the Vandals came? Are you sure he got away?" Susan asked.

"He wasn't captured like me—"

"That's not what I asked," Susan said.

A chill ran through Carolyn. *Did the Vandals kill him?* would be what Susan was thinking. This was a world so different from the one she and Boltzer were used to. Boltzer opened his mouth, then closed it, looking grim.

"What did the Vandals do in Paris? How long did they spend there?" Carolyn asked. Could they have found the missing elly-book, or might it still be there?

"Not long. They scooped up all the lab books from the benches, and then took me somewhere. I was terrified. I wasn't thinking about ..." He trailed off, apparently unsure what he was trying to make an excuse for.

"There might still be this elly-book in Paris. We should go back." Carolyn felt more certain. Her visit there was also cut short, by the French Not-Vandals. Perhaps no one had explored it enough.

"Whoever took *me* had basically set up camp there," Susan said. "If there was an elly-book, they're sure to have found it."

"Oh." Carolyn deflated. "But wait ... if they'd found that, why were they looking for me? It couldn't possibly say *I* had anything, as I didn't exist then." She tapped the book she had open again. "And, hey, they didn't find Ethan's photos. Maybe they haven't searched that well."

"Or maybe they found what they were looking for, but couldn't get into it," said Susan. "Remember, they wanted me—who they thought was you—to open something."

"A genetic lock," said Ethan, animated. "They were all the rage back then. They could have protected their most important research with one."

"But *I* didn't exist yet," Carolyn repeated. "How could they make it openable by me?"

"You existed by the time they got to Paris," said Susan.

"First degree relatives can open them, sometimes further. You can set it. That's why they've gone out of fashion now, what with all the *real* synthetic babies—sorry Carolyn." He gave a quick shrug. "Kids aren't first degree from their parents anymore. Sometimes not even second or third."

Susan reached into her bag and pulled out one of the two black cylinders she had scavenged from the French Not-Vandals. "Is this familiar to you?"

"Oh." Ethan reached out took it. "I haven't seen one of these in decades."

"What is it?" asked Carolyn.

"It's a mitocyl. A quick scan of mitochondrial DNA for identification purposes. You can program it for an individual or just a few SNPs."

Carolyn nodded. SNPs, pronounced 'snips', were Single Nucleotide Polymorphisms, or a single difference in the letters of the DNA alphabet between two individuals.

Ethan continued, "There was a time when law enforcement used them for quick testing of suspects, but it was limited because mitochondrial DNA is only inherited through the mother. Current gene scan devices can sample genomic DNA just as quickly."

Susan took it back. "I'm remembering that now, from our history of policing lessons. Never seen one before."

"Does that mean this is programmed for me? Can we test it out?"

Susan raised her eyebrows. "It stung. I don't think we need to—*we* know who you are!"

"Good point," said Carolyn, still fascinated. "We should go to Paris. Either there's an elly-book there no one's found yet, or the French Not-Vandals found it and can't get at it." She started packing up the lab books. "Susan, you arrange our arrival at the Paris hideaway. We need to get going."

Chapter Twenty-Two

The railway did not seem as run-down as Grampian, but Susan's devices told them most of the train cars had no surveillance at all. There were few fellow travellers. Carolyn and Ethan sat across a table from each other, the pair of elly-books laid out before them, while Susan sprawled on the maroon-and-blue seats across the aisle.

"We didn't want to tunnel from the hideaway, just in case they can be traced, but now should be fine," Susan said.

"That's *amazing*," said Ethan. "Can I? There's so much I want to check on now."

"Me too," said Carolyn. To start with, reading some of her parents' research. But before that … "We really need to BLAST these." She lifted the ID card still around her neck.

"Blast? What?" Susan lifted Carolyn's father's card, which she wore.

"It's bioinformatics," said Carolyn. "We think they're protein sequences. At least this one is." She scooted to the aisle and leaned over to look at Susan's card. As she recalled, it also had short sets of numbers separating longer strings of letters. She squinted. At least at quick glance, they also were missing the key letters which suggested they represented only the twenty natural amino acids.

She showed Ethan how to find the tunnelling app and type in the numbers in the cloud. The now-familiar split screen appeared.

Ethan leaned back with a laugh. "Lynx! My goodness, I haven't seen this in decades!"

"You're familiar with this? It seems to be some kind of text-based web-browser," said Carolyn.

"Exactly," said Ethan. "It was ancient history when even I was young, but the geekiest kids would use it to sneak to the web when our parents refused us smart phones."

"Looks like it's serving the same purpose now," said Susan.

Carolyn typed in the main BLAST website from memory. She was not used to visiting it; she usually wrote scripts that accessed it remotely. Navigating the text-based view was strange, but she finally found the correct place to type in the protein sequence. Typing it in was strange too; she had not actually had to physically type a protein sequence since … well, perhaps forever. Scripts usually generated them from data for her. At the very most she had copied and pasted from one file to another.

So when it came up with only one protein match that was barely a match at all—only scattered amino acids the same across the whole thing—she figured she had typed it wrong. Several iterations later, she was more convinced it was a dud search. Her stomach dropped in disappointment. "I was so sure it was a protein sequence."

"Me too," said Ethan. "Let me try."

"Here, do this one." Susan passed across the other ID card.

Ethan took it and turned an enquiring glance to Carolyn. She shrugged. Her initial excitement had faded. If it was not proteins, what could it be? Even the mystery seemed worthless at this point. She would never figure it out.

Ethan typed away on the touch screen, much more slowly than Carolyn would have. Had she cared about the result, she would have been frustrated at the wait. But the tiny stirrings of impatience barely impacted her mood. The nervous energy that had kept her going for the last few days drained away, and she closed her eyes.

"Turf one!" said Ethan.

She jerked upright and banged her knees on the table. "What?" Had she drifted off? She rubbed her still-smarting knees. She blinked at Ethan, her brain trying to make sense of why he was speaking about grass.

"It's TERF1, the human telomere binding protein."

Oh. A gene, not a word. "Really?" She spun Ethan's elly-book towards her. The match was perfect. One-hundred percent identity across the whole sequence. "How …" What had she done wrong?

She pulled her mother's card off over her head and lay it next to her father's. Why would one have sequences, but the other not? Or maybe only some were … She started typing the next potential protein into her elly-book. Ethan took back the other card and did the same. But despite her faster typing, hers kept coming up blank or with one terrible match; Ethan's hit 100% each time, on more telomere proteins.

Was it something about her interface or her typing? "Let's swap cards."

Ethan pushed her father's card across the table, and she did the same with her mother's. They passed in the middle, and something looked oddly similar between the two. "Wait, stop!" She scanned the pair of cards—that was it! Upside down, they had the same letters, in the same order. She flipped her mother's around, then back. "This one is backwards! Let me try again."

She started from the bottom right of the card. It was mind-bending to type the letters in reverse, and she worked more at Ethan's speed. She was rewarded by a 100% match to TERF1.

Susan crossed the isle to inspect the two cards. "They are identical, even the numbers, just backwards."

"Why would they do that?" Carolyn slumped. "Why telomere proteins? And what do the numbers mean?" They did not seem connected to the genes in any way. One mystery solved, but it did not actually answer anything.

"I think the stuff they were pushing me towards about your chromosomes—or what was meant to be your chromosomes—had something to do with telomeres," Ethan said. "Maybe it has some significance."

"I can't see how we could *tell*." Anger at her dead parents swelled. Why did they scatter these mysteries about? Why did they ever tell anyone *she* had an answer she clearly did not?

"Let's BLAST the rest." Ethan took her father's card—the forwards one—and laboriously typed again.

Carolyn watched him, feeling guilty, then took her mother's and started over, this time typing the first sequence starting from its end. They met in the middle of the two cards, having amassed a collection of more telomere-related proteins.

They lay them out in order—and reverse order. They lined them up with the numbers and other pieces of information about the proteins themselves in various manners. She looked across at Ethan. "Does this mean anything to you?"

He shook his head. "I went into mitochondria."

"So did I." She had not meant to follow her parents' research, but could not she have at least done so in a manner complementary to Ethan? She realised the ridiculousness of that thought.

Susan leaned over. "So just more of a mystery?"

The two scientists nodded glumly. Carolyn tapped on the elly-book. She could use the internet to answer other questions about her parents' work. The stuff they had published, at least.

Chapter Twenty-Three

Carolyn watched as Susan waved her father's ID card past the sensor in the darkness. They had gotten their fill of the internet plus recharging the elly-books during the day in the Paris hideout. She held her breath as nothing happened, but eventually the door popped outwards as it had for her before.

They followed Carolyn's lonely footprints through the dust to the stairwell. "How did you get in the first time?" asked Carolyn.

"The Not-Vandals used the front door. They had a card already," said Susan.

"So did the reporter," said Ethan. "Or I suppose that would be the not-reporter." He held the stairwell door for Carolyn. "Your people didn't use the lift, did they?"

"No," said Susan.

He looked longingly at the smooth metal doors. "Probably can't trust it then."

Carolyn sympathised. Her thighs burned as she climbed, muscles sore from her journey up the Bucharest stairs—three days ago, was it? She was losing track of time. It might actually be a full week since everything had started by now. How were her intro biology students doing, she wondered. Who would mark the essays? Surely the University had figured it out. But what about Frank's PhD? What if this whole series of events delayed him too much—they would give him more time to deal with death of a supervisor, right? She pushed down panic, before it could hit the target she knew it was aiming for: Ellen. Keith and Bae would let her know her mother was not dead. And some good soul in the police—or whatever part of the police was in league with the

Vandals—was protecting her existence from them. Ellen would be fine. She really wanted to believe that.

* * *

Carolyn sunk to the floor. Ethan sat in her father's desk chair. Susan appeared in the doorway and shook her head. Nothing of interest revealed in a second search of her mother's office, either.

Nothing. Ethan and Susan had been right. The Vandals and Not-Vandals had picked the place clean, if anything had been there at all. Signs of the Not-Vandals were gone as well, suggesting they had moved on to the next step, whatever that was.

"Back to the tunnels? Figure out what to do next?" Despair tugged at Carolyn, but she refused to let it take hold. They were so much further ahead than the last time they were in Paris. They knew there was really something to find. They knew there were at least two—possibly three—other groups after the same material. But they were the group with Ethan on their side.

They retraced their trek down the stairwell to the dusty corridor. A mess of scuffed areas now covered Carolyn's first, lonely prints. Susan took the lead towards the outside door, and Carolyn looked behind them. The scuff marks extended well past the stairwell entrance.

Her mind took a moment to process the information. Someone else had come after them! She turned to say something to the others. But Susan was already at the door: she opened it to reveal three men and a woman in the alley.

"You!" said Ethan. "And *you!*"

Chapter Twenty-Four

"You! And *you!*" echoed Susan.

"Who?" Carolyn's legs itched to flee. But while their voices sounded shocked, neither of Ethan nor Susan appeared ready to run.

"You *kidnapped* me!" said Susan.

"You *lied* to me!" said Ethan.

Then the two of them pointed at the remaining pair of people and chorused, "And what are *you* doing with *him*?"

The second set of *you*'s held up their hands. "Let's get somewhere we can talk about this calmly," said the woman. She had a heavy French accent.

"Calm? *Calm?*" Susan leaned toward the woman, fists clenched tight at her sides. Carolyn recognised the air of hysteria in Susan's voice, so similar to the tone in another alley, a lifetime ago. "Did you set us up? I thought it was too easy."

"Look, this isn't the place," said one of the men. He sounded American.

"What, precisely, *is* the place, Agent Dorsey?" asked Ethan.

Carolyn's mind spun. Susan and Ethan recognised these people, but she had no idea who they were. 'Agent Dorsey': could that be one of Ethan's FBI agents? What would he be doing in Paris? She supposed that was exactly what Ethan and Susan had just asked.

"Let's all go back to the cafe," said another man, his accent French. "We can discuss things there." They backed away from the open door, and Ethan and Susan passed through.

"Hold on." Carolyn stopped in the alley, door still open, propped on her shoulder. "Who are you? Why should we go with you? How can we trust you?"

"Let's not argue here," said Agent Dorsey. "*You* have caused a great deal of trouble, Dr Gray. Or should I say Schwarz?"

"Gray. *I've* caused trouble?" Her voice rose in indignation. "If everyone hadn't been so keen on being mysterious and just *asked* me outright, 'Oh, hey, do you happen to have the Schwarz Final findings?', I would have laughed and said, 'No, that's a myth' and people could go on their way without destroying my life!" Her fists were clenched, nails digging into her palms. She stood on tiptoe, towering over everyone except Agent Dorsey.

"It's a myth?" asked the man who had not yet spoken, his accent also American.

"No," said Ethan tiredly. "At least we—"

"As far as I was concerned it was," interrupted Carolyn, her anger still high. "I don't care what they may or may not have found. I just want my life back." Although as she said it, she knew inside she *did* care what they had found. What could be so amazing that people were searching for it some thirty-odd years later?

"Me too," said Ethan.

"That makes three of us," said Susan, her voice also fatigued. She scowled at Ethan. "At least you haven't had yourself supposedly killed off. You've got something to go back to."

"Um," said the second American, "that's not quite true."

"I'm dead too?" asked Ethan.

"Really not the place," said Agent Dorsey. "Let's *move*, before someone comes upon this little reunion."

The four of them started walking. Carolyn, Susan, and Ethan shared glances, then followed glumly. Carolyn supposed it was not any more absurd that anything that had happened so far, although she wished she knew as much as the other two about their new companions.

*   *   *

They settled into a round, tall-backed booth in the back of the 'cafe', which seemed more like a nightclub than a cafe to Carolyn. The decor was dark, leather and wood, accentuated by dim lights. Thumping music filtered in from the front room, where the floor had exuded the smell of beer. Perhaps with a good mopping, different lighting, and some flowers on the tables in the daytime, it could be cafe-like.

Susan spent a short time investigating the devices in her pockets, then sat, appearing satisfied, if wary. Ethan slumped heavily in the booth on the opposite edge, and Carolyn perched on the other end of the circle next to Susan. At least their mysterious companions had voluntarily slid to the middle of the circle, leaving the edge for the three of them. Not being blocked in made Carolyn feel better.

"Okay, who *are* you?" Carolyn asked.

"Agent Dorsey, FBI," said the tall American. "I'm lead investigator on the cold case of your parents' murder."

Murder. The word thumped into her heart harder than she expected. She had already heard speculations from Ethan. But stark words from an investigator seemed to carry more weight. "You're sure they were murdered?" Her voice squeaked, and she winced. She was getting distracted, but she suddenly, absolutely, had to know.

Agent Dorsey nodded. "Positive. The question now is by whom."

"I—" She bit back more questions. "Who is everyone else?"

"This is Twelve." Susan gestured at the woman. "She's my contact in the hideaway. What I want to know is what you're doing with *him*." She pointed to the French man beside her.

"And *I* want to know who *you* really are." Ethan squinted at the other American. "Pretty sure you're not a journalist."

The man shook his head. "Sorry about that. But I wanted to get you away from *them*,"—he gestured with his head towards Agent Dorsey—"without trouble."

"You *should* have gone through proper channels!" said Agent Dorsey.

"I tried," said the other American. "But your boss—"

"*My* boss? How about *your* boss?"

"Okay, both our bosses were obstructive. But there was stuff I needed to know from Dr Boltzer."

Carolyn's frustration built. She half rose from her seat. "This is all very fine, and I'm sure you've got some things to work out, but would someone please tell me *who these two people are!*" She thrust both arms out, pointing, to the two unidentified men.

"Spies," said Twelve.

"I'm not—" started the French man.

"The term is—" the not-journalist said.

"Spies," said Agent Dorsey levelly, cutting them both off.

The spy not journalist heaved a big sigh. "Intelligence agent is preferred," he muttered.

"Not corpsies?" asked Carolyn, voice rising.

"See, that's why we prefer intelligence agent," said the American spy. "No. Not corporation spies. Good, old-fashioned government intelligence."

"Oxymoron," said Ethan.

Agent Dorsey raised his eyebrows. "Sometimes I think so," he agreed.

"So which governments? What was the purpose of kidnapping me?" Susan demanded.

"We thought you were her." The French spy gestured towards Carolyn.

Carolyn directed an incredulous stare at the small, blonde Susan, her near opposite. "How could you possibly—"

"She might have taken after your father." The spy sounded defensive. Perhaps she was not the only to question the lack of similarity.

"So what would be the purpose of kidnapping me?" asked Carolyn.

"Because spies have an allergy to telling the truth," said Twelve, with reproach in her tone. "They told *us* they were going to talk with you, not swoop and nab."

"Could somebody just start at the beginning and explain?" shouted Carolyn, fed up with the bickering. She found she was half standing. She slowly sat back down.

The American spy raised a hand, as if in a classroom. "Our French colleagues here obtained access to Vivcor Paris and realised the place had been abandoned *in-situ*, as it were, meaning it could shed light on whatever was happening at Vivcor at the time of the Schwarzes' deaths." He clasped his hands together and laid them on the table. "They found several items which appeared to be in some kind of code—which is how I got involved—and also wanted some scientific translation of the Schwarzes' material. The original plan would be I'd come over, deal with the code while the French copied all the science in the building. Then I'd take everything back to the States and consult with Dr Boltzer, who was already cooperating with my friends at the FBI. But I was nearly on my way when it became apparent that Sandslin had discovered the same thing, so I cut corners to get Dr Boltzer to come with me."

Sandslin. Her funder that Carolyn had spoken to. Who had mentioned the Schwarz Final Findings and made up some false story about her parents being at LSU. Could *they* be the Vandals? She stored that hypothesis away and focussed on the rest of the American's story.

"Turns out we didn't cut corners close enough." He turned to Ethan. "Really sorry, it was a bad call on my part, involving a civilian like that. But I'm glad to see you alive and well." He rested clasped hands on the table. "Even though Sandslin made off with all their research."

Ethan opened his mouth, but Carolyn placed a hand on his knee under the table. She wanted to hear the rest of their story before revealing that they had the photos from the Paris lab, plus the Bucharest books.

"But while my European friends were off chasing you—something they were well better able to do than me,"—he looked contritely at his hands—"I was able to break the code that Frenchy here found, and it led us to a gene-locked safe." He raised his face to Carolyn. "We hoped you would be able to open it." He spread his hands, indicating the end of the story.

"That doesn't explain why you grabbed me like that," said Susan.

"Poor judgement." Agent Dorsey glared at the French spy.

"A mistrust of civilians." The American spy gave an almost understanding shrug.

"We couldn't be sure she wasn't in league with Sandslin," said the French spy, finally speaking up for himself.

"Using *our* network?" Twelve's voice was incredulous.

"She takes money from them. That's on record." He crossed his arms. "Then here she was in Paris, despite having died just the day before. That smacks highly of a corporate touch."

"Yes!" Susan smacked her forehead lightly with her hand. "To discredit and create problems for her, just like they did for me. Not to … to …" She trailed off, apparently unable to guess what the Frenchman thought Carolyn might have been doing.

"To enable her to travel freely in her parents' footsteps, with governments none the wiser," supplied the French spy.

Carolyn creased her forehead. Were governments and corporations really such enemies as this? Governments used corporations for everything nowadays, even law enforcement, like Susan's employers. She felt the same vertiginous sense she had when contemplating the thought that the criminals might own the police. Where were the good guys in this, anyway? *Were* there any?

"That's clearly not the case." Susan patted Carolyn's elbow. "All we want is our lives back."

Ethan's and Carolyn's gaze met across the table. "And to find out if my parents really *did* have secret findings," said Carolyn. Because it was true: the mystery had been raised. Yes, she wanted to fix her life and make sure Ellen was safe, first and foremost. But a close second came digging to the bottom of this mystery. The look in Ethan's eyes suggested he felt the same. What if her parents really had discovered some world-changing result? It was something they had wanted released, eventually, if what Ethan said about their hinting was true.

"So you'll help us open the safe?" asked the American spy.

"Of course!" She half stood, ready to go.

"Not right now." He waved her down. "Getting there is going to take some planning."

"We've seen it," said the Frenchman, "but haven't yet gotten close. We'll have just one shot at it, as getting to it will alert Sandslin."

"Are Sandslin the Vandals?" Carolyn asked.

"You mean the people who kidnapped Ethan and broke into your lab?" asked Agent Dorsey. "Yes, that's them."

"Okay, and *you're* group number three, that was hanging out in my parents lab and kidnapped Susan." She looked at the French spy. "You're all working together?" She waved her hands vaguely at the four of them.

"We're just helping." Twelve cast an unreadable glance at the French spy. "And staying off the corporate radar."

"Of course," said the Frenchman in a reassuring tone.

"Do you spies—sorry, intelligence agents—have names?" Carolyn was tired of thinking of them by their accents.

The two spies looked at the table.

"He told me his name was Mike Talon," said Ethan.

"Mike Hafal," Mike muttered. "Call me Mike."

"Call me Albert," said the French spy, pronouncing it like 'al-berre'.

"Okay. So Agent Dorsey, Mike, and Albert here are group three. But who was it that came to my Uncle's place in Scotland. Was it you?"

The three men looked at each other. "I know nothing of this," said Mike slowly.

"They were there the same time my 'death' was discovered down in London," said Carolyn. "They were looking for my parents' stuff and specifically asked after ID cards."

"They could still be Sandslin," said Mike. "It's not like those with Ethan were all they have."

"But they wouldn't need ID cards," said Ethan. "They got into Vivcor Paris just as easily as you did."

Carolyn grasped the card on her chest. "But the cards aren't just for access." She swivelled to face Mike. "You said something about a code being why you were here. Why's that?"

"I'm a cryptographer. I break codes for a living." He waved his hands in the air. "All this running about isn't my kind of intelligence."

These four seemed to be forthcoming. Susan had trusted Twelve; perhaps to her detriment, but Mike and his friends had not actually harmed Susan. She and Susan probably had done more damage to them in their escape. Carolyn came to a decision. She pulled her mother's ID off her neck. "We've got another code for you."

Chapter Twenty-Five

Mike reached out and took the card. He brought it closer to his face and lifted his steel-rimmed glasses. Near-sighted and aging: Carolyn had seen the same gesture in her slightly older colleagues.

She pulled out her elly-book—the student's lost one—and folded her arms over it. "The letters are protein sequences. We don't know what the numbers are. There are two cards—this one and my father's. The only difference is that this one is backwards: the sequences read from the bottom right." She opened the elly-book, brought up their notes from the train, and pushed it across the table towards Mike. "I can't imagine there's a code in the protein sequences themselves, since that's biology not cryptography, but maybe in their identity? We've identified all the proteins. They're all human, and all have something to do with telomeres."

"What's this?" Mike ran his finger down the columns of Carolyn's scribbled notes.

"Each protein has a variety of information that could be connected to it. First its name: they've got a short name, like this one here, TERF1." She leaned across the table to tap at the screen. "That's an abbreviation for the longer name here." She hovered her finger above the words, *telomeric repeat binding factor 1*. "Each protein also has a bunch of accession numbers for different databases, like UniProt, Ensembl, and so on." There had been talk for ages of some kind of unifying system, but it only got worse, as new databases appeared with their own systems. "Then there's information about the protein itself. Like what chromosome it's on, where it starts, whether the initial methionine is cleaved or not, and how much of the protein was written down. All of these start at the beginning of the protein, but obviously don't have the whole thing." She gave a little laugh, but could see that only Ethan understood. Proteins had hundreds of amino acids in them, far more than were printed on the small card. Each cut off at a methionine as well, making the confusion for the reverse card.

"Is this all the information?" Mike asked.

"All we could think of. We could not see anything obvious." But none of she, Ethan, nor Susan had any experience in codes.

Mike pulled a sheet of docfilm from somewhere and held it above the elly-book. "Mind if I?"

"No," she said cautiously, not quite sure what he was asking.

Mike laid the docfilm on the elly-book and tapped a short pattern in its upper left corner. The docfilm flashed pink, then held the same image as on the elly-book. Her jaw dropped open. She snapped it shut. Susan had shown her docfilm could be hacked, and now she saw elly-books could be copied. Neither of the things science relied on for its immutable record or its privacy actually served their purpose.

"Is there information about codons?" asked Mike.

"Codons?" Her brain had trouble processing the question, still stunned at her revelations. "No, they're already protein sequences."

"Oh, just that's the one thing I know about proteins, or biology really. The codon code." He smiled disarmingly. "It's an example of nature's code. How genes are made up of three-letter codons, each of which corresponds to a particular amino acid in the protein. And how it's a degenerate code: more than one codon corresponds to each amino acid."

"Yes, except for methionine and tryptophan." Carolyn slid into teacher mode. "The others all have multiple codons, even stop: there are three codons which don't mean any amino acid, just correspond to the end of the protein. They're present in different frequencies; it's one of the ways cells control expression of genes."

"Oh?" Mike sounded honestly interested. The rest of their companions seemed to fade into the background, leaving only Mike's eager, curious face gazing at her across the table.

"Yes, some codons are more 'common' than others. Each codon corresponds to what's known as a transfer-RNA, or tRNA: this has the complementary sequence that binds to the codon on one end and the amino acid on the other. Some tRNAs are more common than others, so the more common the tRNA for that codon is, the faster the gene is translated into a protein. Proteins that are needed very fast tend to have high-frequency codons. It's known as optimal-codon usage. We use it in the lab when moving a gene from one organism to another. For example, if you put a human gene into a yeast cell, the tRNA frequencies in yeast are different than in humans. It might turn out one of the ones used a lot in humans is really rare in yeast, so your protein takes a long time to make. So we optimise the codon usage before putting the DNA in."

He opened his mouth, as if to ask more, then looked at their companions. "I didn't entirely follow all that, but it sounds fascinating. Maybe you could explain more later?"

She blushed. "Sure." She had gotten a bit carried away. Teacher mode had taken over in the face of Mike's apparent interest. She felt flushed with the excitement of talking about her science, and perhaps with something else. She looked away from Mike, avoiding the thought. "When can we visit the safe?" she asked instead.

Mike turned his head towards Albert, and Albert said, "First thing tomorrow. We should get there early in the morning." He tilted his head. "Do we need Dr Gray?"

"I thought that was the whole point," said Susan.

"I think he means just on the trip," said Mike. "We'll be heading out into the Paris underground, beyond Twelve here and her friends' domain. It'll be treacherous. Better to have just me and Albert if possible."

"And me," said Agent Dorsey, his voice nearly a growl.

Mike shot Agent Dorsey a look, as if to say he would address that later, and turned to Albert. "Approaching it could tip off Sandslin to its location. If it does need a live person to open it, we'll have ruined our chance."

"Um, what do you mean by 'live person'?" Carolyn had been reasonably confident these people were on their side, and Mike particularly seemed rather engaging, but that was an ominous statement.

"As opposed to just a tissue sample," said Mike with a laugh. "Not as opposed to a dead person, of course! Some of these worked with a blood sample or even a few cheek cells. But there's no way to tell until we get there, so you'll have to come with us."

"You'd take Carolyn on her own?" said Susan, protectively.

Agent Dorsey leaned back and crossed his arms. "The more people on the trip, the more dangerous it is. You and Professor Boltzer aren't necessary, so you shouldn't come." He patted Ethan's shoulder, sitting beside him. "It's really not a journey for the Professor. The old tunnels are a tight squeeze at times, and it'll be close to a ten kilometre walk with all the winding."

"That'll take hours!" said Carolyn.

Mike nodded. "It will. Dorsey is right that the fewer the better." He cleared his throat obviously.

"I'm on strict instructions to not let you out of my sight again," said Agent Dorsey. "I'm definitely not letting you do this on your own! The Bureau would have my ass." He made a throat-clearing noise, more natural than Mike's. "Again."

Mike sighed. "Right. I hear you. It'll be us four."

"Carolyn hasn't agreed to this yet!" Susan gestured towards Carolyn.

Carolyn took Susan's outstretched hand in hers and squeezed. Susan returned the squeeze, her soft, warm grip giving Carolyn confidence. "I'll be fine," Carolyn said. Ethan knew Agent Dorsey and Mike; he trusted them. Look where that had gotten him. She dismissed the thought. She was committed now. A gene-locked safe, which she could open: that must have been what her parents meant when they told Ethan to look to her for the answer. This *had* to have the secret.

Chapter Twenty-Six

The five of them ate a very early breakfast in a small cafeteria set in a bulge protruding from a tunnel, with silver folding tables and chairs. Carolyn's and Susan's attempts to get the agents to discuss today's destination fell flat; too many cars around, apparently—protecting the listeners as much as the speakers. Carolyn grew uncomfortable with the lengthening silence. Her mind turned back to her question of the day before. "Are governments and corporations really so much at odds?" she asked. "I mean, how does that even work? Aren't the police subsids?"

Susan bristled beside her. "The police are contracted by subsids; but police officers aren't corporate."

"No, I mean—" *When I was a kid, cop didn't mean corporate*, Carolyn recalled Susan's impassioned words. She didn't mean to offend her friend.

"Law enforcement works for the people, as we always have," said Dorsey. "It's just that now ..."

"Now we also work for someone else. Technically," Susan said, apparently past her brief upset. "Like science—you have corporate funding."

"True." Carolyn remembered Bae's gasp at Sandslin being her funder. Sandslin: the Vandals. She shivered. "Does the FBI have corporate ..." *Overlords* wouldn't exactly be the most politic thing to say.

"The FBI is still a wholly federal institute," said Dorsey.

"With all the benefits and inefficiencies that brings with it." Mike held up a hand as Dorsey leaned forward. "I put my crowd in the same boat! We might work better if we did have corporate connections."

"No," said Susan. "It doesn't help. Yeah, you get more data, more access. But sometimes … it just seems like there is a lot we don't know. You never quite know why—and if somebody does know and just isn't telling you."

Dorsey nodded knowingly. Carolyn thought back to her other revelation: if the corporations owned the police, and even managed to keep groups like the FBI in the dark, who *did* know what was going on? Because clearly, corporations didn't know about places like they sat right now; otherwise they wouldn't exist. But Susan had said the police did know about the communities. The police knew; maybe their *employers* didn't. Was it better or worse to think that everyone was hiding half their information from everyone else?

She looked past Mike to a family settling down at the table nearest the corridor. A boy Ellen's age was laughing at some joke made by his younger brother, who sported a satisfied grin. Better, she decided.

Soon after they finished, Albert showed up, accompanied by another Frenchman. "I am Twelve," said the newcomer.

Carolyn furrowed her brow and turned to Susan. "I thought the woman last night was Twelve," she whispered.

Susan whispered back, "It's a position, not a name. Twelve is the contact for outsiders. They don't even know each other's names, I think."

"Follow me." The new Twelve started off at a brisk pace down the tunnel.

It was a more abrupt start than Carolyn had envisaged. She stood and gave Susan a hug.

The smaller woman stiffened briefly before returning it. "Good luck."

"Be safe," said Ethan.

The FBI agent and two spies had already started after Twelve. Mike looked back and gestured.

"Thanks. See you later today with the answer!" She grinned, then trotted to catch up. Her mind spun. The sudden departure, then being nearly left behind, was disorienting. The fact that she was now on her own with people only met yesterday—or now, in the case of the new Twelve—slowly sunk in.

They walked briskly, in silence. She wished someone would say something, but she was too intimidated to start a conversation. What seemed like nearly an hour passed. The decor around them changed, from the smooth, white walls that had given the place the sense of a science-fictional space station, to grimier and grimier white tiles, eventually to grey concrete. The tunnel narrowed, such that the five of them walked one behind the other. She found herself nearly at the back, with only Mike behind her. The light dimmed as well, until a thin strip of LEDs tacked to the ceiling was the only illumination.

"This is as far as I go," said Twelve's voice from the front.

"Thank you," said Albert. There was a crash of clanking, then rusty creaks. "Take these." The latter appeared to be directed to Agent Dorsey, as a moment later, he handed Carolyn two torches, keeping one for himself. She passed one back to Mike.

"Um, where are we going?" She stared nervously at the torch. They had said it would be underground, but she assumed it was more of the space-age tunnels they had been staying in.

"To the safe." Albert's voice was muffled and came from the right.

Agent Dorsey vanished into the wall. She could now see an open grate to her right, topping a spiral metal stair, encrusted with rust. It wiggled with the weight of Albert and Agent Dorsey descending.

She paused at the doorway. "Are you sure this is safe?"

"This isn't the safe," said Albert, "it's the way *to* the safe."

"No, I meant, will this stair hold us all?" Carolyn smiled at the small misunderstanding, but her humour was muted at the thought of descending further, into some dark tunnel. Give her balancing on the outside of a building any day over this. She should have questioned the exact path more back at the cafe, when the others had been forthcoming. What exactly did the 'old tunnels' mean?

The stair stopped wiggling, and Agent Dorsey's voice floated up. "It held the two of us fine, come along."

She still hesitated. "Where are we going? Into the catacombs or something?"

"Only just a little bit," said Mike.

She snapped her head back towards him. "Only just a little bit *what*?" Her voice squeaked, and she winced, although unsure why. It was perfectly acceptable to be nervous about descending into this unknown darkness. She had meant the catacombs comment to be a joke; it appeared perhaps closer to the mark than she had hoped. She stepped cautiously onto the stair.

"We only go through the catacombs a little bit. The safe is in more modern tunnels, not as new as those above us, but probably made by cataphiles."

"Who?" She kept one hand on the central pillar, another on the railing to her left.

"Cataphiles." He followed her onto the stairs. "The predecessors of our friends above, one might say. The name means lover of the catacombs. They were a movement in the late twentieth century and reasonably far into the twenty-first, part of a larger movement of 'underground explorers' who enjoyed finding what they considered hidden treasures underneath the ground." She descended, concentrating on breathing carefully as Mike continued to talk behind her. "Some of their finds were quite fascinating, actually, like bunkers from both sides of Word War Two, very near each other but clearly in complete ignorance of the enemy's location."

She reached the bottom, her worries somehow calmed by attention to Mike's story. She played her torch about her. They were in a larger tunnel than the one above, with a different colour scheme: light tan of the natural rock. The temperature was the same, but the tunnel smelled old, like a disused library. "You said you've seen the safe—you've come this way before, then?" Now that he had stopped talking, her fear rose again.

"Both of us have." Mike stepped forward and gestured at Albert.

She edged closer to Mike. "But you said you think the tunnels were *made* by the cataphiles? So they weren't just explorers?"

"Some only explored," said Mike, taking the lead. She skipped to get close behind him. The raw edges of the tunnel seemed less secure than those above, and her heart thumped heavily. "But, especially later on, as the corporations gained more power and it became useful to have places to hide, they did their own excavations. There were even small businesses: nightclubs, movie theatres, and so on. As I said, the predecessors to our hosts above. It was more …" He tilted his head, as if searching for the right word. "More hedonistic than currently. Everyone was an individual, and they were doing it for fun."

"You sound like you'd like to have lived then," she said, surprising herself by voicing the thought out loud.

He glanced back at her. "I guess I do, a bit. It seems like it was a world of more possibilities. Time for wandering about illegally underground, for little reason other than you wanted to."

"As opposed to needing to." She was not particularly happy about wandering about underground. If it was for fun, not on the trail of lost secrets of her parents, in a race with some corporation—Sandslin, that would be; she still had trouble processing that one of her funders could be the Vandals—would she be happier about it? She ducked as the ceiling got lower. She did not think so.

∗   ∗   ∗

Carolyn and Agent Dorsey hunched over, while Albert stood easily and Mike's head was fingerwidths from brushing the ceiling. Mike said, "It's just around the next bend. Albert and I will go ahead to make sure the way is clear, then come back for you two." The two spies vanished around the corner.

She leaned against the wall and exhaled a big breath. They were almost there. Mike had pointed out the period, just shortly before, when they passed through the ancient catacombs, but it seemed little different. No piles of skulls or corridors full of bones like she had imagined. She squinted at Agent Dorsey, finding herself wishing that it was Mike who had stayed behind with her.

She sat, relieving her back of the strain of standing bent. Her legs tingled; she had walked more, and over tougher ground, than she thought she may ever have. She rubbed a calf. She wondered how much time had passed. It felt ageless down here.

Footsteps sounded, soon followed by a bobbing light. It was Mike. "Come along." He reached out a hand to help her rise. He put a finger to his lips. "Be quiet. We don't know what recording devices might be about."

She hugged her chest in her hunched-over walk. Her nerves were rising.

The tunnel got shorter, until everyone bent nearly double. Ahead, Albert sat silhouetted at a bright opening.

Mike approached the opening at a crawl. He pointed at Carolyn, then gestured as if to say 'follow'. She sunk to her knees and crawled after. The room beyond took shape. It was grey concrete, what she was starting to think of as the 'mid-age' tunnels. Light came from strips of LEDs again, this time strung in loops along the walls. It was large, with a tall ceiling that reached at least two stories up. The opening out of which they peered was placed approximately a meter and a half above the floor. Two normal-sized doorways broke the walls to the left and right, showing dark corridors beyond.

And the room itself. It was like some kind of junkyard. Objects piled haphazardly, reaching metres up near the walls, and gathered in shorter stacks in the middle. Paths wandered through, splitting and re-joining like the flow of water on a mud-flat as the tide receded.

"The safe is there," Mike whispered, pointing towards a pile on the far wall.

"I don't see …" Then it popped out at her: a dark grey box, very safe-like, with a big lever in the top left. A tarp draped over half of it, but left uncovered a protruding square with a stylised double-helix symbol. "Why in the world is it *here*? What is this place?"

"It would have taken some effort to get it here; those things are heavy. Assuming this dumping ground has been like this for the past few decades, I presume your parents picked it because one more item would not stick out," Mike said. "We're lucky it's visible. Otherwise finding it would have been much harder. Entering to explore could have tipped people off."

"How did my parents know about this room?" The oddity of whole thing suddenly struck her. A safe in a junkyard beneath Paris. "Does this mean they were cataphiles?"

"Perhaps," Mike replied. Albert put his finger to his lips, shooting a glare at Mike. "Right," Mike said. "The plan is I'll go down first, then Albert will help you after me. You two watch those doorways, and we'll work the safe. Ready?"

"Ready," whispered Agent Dorsey. Albert nodded.

Mike swung his legs over the edge, rolled onto his stomach, and dropped lightly to the ground. He froze for a moment, looked about, then gestured up. Carolyn copied his motion, and Albert held her shoulders, then arms, as she dropped down. Mike steadied her on the ground. She and Mike wound quickly towards the safe. She kept her eyes on Mike's back ahead of her, trying not to be distracted by the objects around her. Was that an autoclave? And that looked like the front seat of an automobile. She snapped her eyes back to Mike.

The safe sat on top of what appeared to be clothes washers, which were slowly crumpling under its weight. Mike stacked two tyres in front of the washers, and they stood on top, with the safe at chest level.

"How do we open it?" Her heart thumped. She was mere inches from her parents' secret.

Mike pushed the tarp aside, revealing several recessed panels. The tarp cracked and powdered along the edges. He tapped at the recessed panels, generating plumes of dust, until one swung back. "It looks like we did need you. Good thing you came. Put your finger here. It'll probably prick, like the mitocyl."

She lifted her hand and hesitated. "Albert and his friends used the mitocyl on Susan, not me. She said it hurt."

He shook his head. "It's just a quick prick, sampling skin cells. I would expect, at least …" His voice trailed off.

She squinted at the safe. "What if it's going for blood? How can we even know it's clean?"

He tensed beside her. "Sandslin could be watching us right now!" he said sharply. Then he visibly relaxed. He squeezed his eyes shut.

He opened them and patted her arm. "You've come this far." His voice was soft, soothing. "We're in a lot more danger delaying here than using this would probably be."

She was unsure about his estimate of the relative dangers of blood-borne diseases, but his calming voice, and the undeniable fact that she had walked for hours through uncomfortable tunnels just to get here, reminded her of her purpose. Before she could change her mind, she shoved her left index finger in.

She felt heat, then a sharp pain, which faded. Some lights on one of the recessed panels blinked blue.

"Hmm." His face screwed in concentration. "Are you sure you're Carolyn Schwarz?"

"Of course!" She was not entirely sure whether or not he was joking. "What's wrong?"

"It should give some more positive signal. We'd hear the unlocking. Try again."

She did not want to, but she plunged ahead before thinking too much, like last time. The same heat, pain, and blue lights repeated.

Her heart sank. Maybe it was the wrong gene-locked safe. Maybe it wasn't set to first-degree relatives, but only her parents. Yet … "What about just trying the handle?" She reached out and pushed down. The handle swung easily. Four sharp clanks, and the door opened.

Chapter Twenty-Seven

The safe was empty.

Carolyn's heart sank. It couldn't be! She reached in and patted around, but her eyes had not deceived her. It was empty. It had not even been locked.

She turned to look at Mike. His face showed the same disappointed shock.

"Could they have gotten here before us?" she asked. "No," she answered herself, "there was all that dust." The safe had not been disturbed.

"Why leave such a complicated coded message?" asked Mike, clearly following his own train of thought. "Maybe …"

Loud bangs, like shots, echoed through the room. She turned right, towards the sound, and saw Agent Dorsey diving beside the door.

"Run!" Mike grabbed her hand and jumped down from the tyres. She ran alongside, half dragged by his arm, towards the door that Albert had been guarding. Albert passed them, running back to Agent Dorsey and holding a gun in his right hand. They were past him in a moment, but the sight of the gun seared into her brain. She could still see it, a passing glimpse, held as casually as any action hero on the 3D.

Terror and a strange sense of disconnect mingled. This was not her life. She did not run about with people *shooting* at her. More shots sounded; Agent Dorsey and Albert must be shooting back.

She and Mike ran down the corridor, lit by Mike's torch. Where was hers? Sudden panic rose. She could not remember. Had she put it down near the safe? Her hand in Mike's sweated. She gripped tighter, terrified of losing him and being lost, alone, in the dark.

Mike dove left, then right, then left again. Tan walls bobbed past in the torch light. They were in the old tunnels again. Was that a bone? Her mouth felt dry. She did not want to be running underneath Paris through a grave-yard. Mike stopped.

She stumbled into his back. "Wha …?"

"Sush." He put a finger to his lips.

The only sounds were their joint breaths, echoing in the close corridor. His eyes shifted side to side, glinting in the torchlight.

"Quietly now," he whispered, gesturing forward. They walked slowly, in silence. He stopped every so often and listened. Eventually the tunnel they were in came to an intersection, with another tunnel at an angle. Large rocks lay at the join, as if they had fallen from above. He played his light up. The walls rose into darkness; they narrowed, but the beam did not reach the ceiling. "Wait here." He released her hand.

She grabbed it back. "In the dark?" Her voice squeaked, and she did not care. She would have rather shrieked.

"Where's your torch?"

"I don't know." Tears welled. Losing the torch suddenly seemed the worst thing in the world, as if she were a small child admitting to breaking her mother's favourite cup. He would think her a silly incompetent.

He frowned. "All right, come along."

She followed him down the tunnel that went off to the right. The ceiling got lower until they both crawled. Then it dropped further. He lay on his stomach and scooted forward.

"Um … is this safe?" she asked. But she was thinking, *No way in hell am I doing that.*

He stopped scooting. His voice filtered back, muffled. "Wait there. I'm not sure if this will work."

She dropped to her elbows and drew her knees to her chest, curled in a strange, squatting foetal position. The rough stone's sharp edges dug into her elbows and knees. Terror at being left alone warred with terror at the narrow passage ahead. Her breathing echoed in the close tunnel. The tunnel ahead got dimmer and dimmer. Then she was in the dark. She closed her eyes.

It was no different. She snapped them open again. "Mike?" No response. Panic gripped her chest. "Mike!" she shrieked. The sound echoed, *ike-ike-ike.*

"Sush!" came back a faint word. Had she imagined it?

She did not yell again. She closed her eyes, pretending the darkness was from her closed lids. Not the stygian blackness of the catacombs. She imagined she could hear noises. Crunching. Creaking. Crumbling.

Rats. *Just rats*, she told herself. Cute rats like the ones in the animal house at the University. Not skeletons spontaneously forming and coming after her. She berated herself for the immature, unscientific fear. But now that she had the vision in her mind, imaginary bones rained off the walls around her. She opened her hand on the cold rock and scrapped it with her fingernails, feeling its solidity.

She rested her forehead on the back of her hands. *Rats*, she said to herself desperately. *Mike, come back soon!*

*   *   *

Scrapes, in more than just her imagination. She jerked up and banged her head on the celling. "Ow!"

"Carolyn?" It was Mike's voice.

"Mike?" Relief flooded. She started sobbing. Faint bluish light filled the tunnel ahead. She could not stop crying. She sniffed, wiped her nose, and curled into a full foetal position on her side. Soft hands patted her shoulders, and a thread of embarrassment wound through her. Mike must think her a coward. But the release of pent up terror was too much. She kept crying.

"I'm back. It's okay. I'm back." He repeated his words more times than she could count, patting gently.

Finally, she stopped crying. Her cheeks heated. "I'm sorry." She pushed herself back to all fours. The tunnel was still lit by the strange bluish light. She finally sourced it to a mobile phone in Mike's hand. "Where's your torch?"

"Batteries ran out." He rattled it in his hands.

"You mean we're stuck down here in the dark?" Panic began to resurface.

"Not the dark." He waved about his phone.

"So, um, is that the way out?" Her voice squeaked, and she was embarrassed again. Amazing that she could be after her little sob-fest.

"Afraid not."

"We're not lost, are we?" They were going *die* here, under Paris, and frighten some other poor lost soul as a pile of bones.

"Not exactly." He cleared his throat. "Maybe a little."

Chapter Twenty-Eight

Irrational anger seized her. She wanted to reach out and strangle Mike. "Maybe a little lost!?" She had trusted him. She had trusted all of them. Perhaps she should have taken her chances with Sandslin rather than racing off into the tunnels.

"I've gotten to the junk room this way—or what I thought was this way— before. There's a super-narrow spot like that." He gestured at the tunnel from which he had just emerged. "But that wasn't it." He wiped his sleeve on his chest. "Either that, or the way's flooded."

"So, we just go back?" Her heart seemed to beat in her ears.

He nodded and scooted past her. She crawled after him. It was a relief when the ceiling raised enough they could stand. She stretched. "We don't have to be quiet anymore, right?"

"I guess not."

"Because at this point I'd rather be found and taken out of these tunnels," she said. She was not sure if it was a joke.

"Or shot," said Mike. She was not sure if that was a joke, either.

The intersection was a welcome, familiar sight, if dimmer in the blue light of Mike's phone. This place was navigable. They would not be stuck down here forever. Mike headed down a tunnel angling off to the left.

Carolyn turned after him. She stopped. "Wait." She backed up. "That was more of angle than I remember." She thought they had turned very obliquely to the entrance, not that sharp angle they had just taken.

He came back and played the phone about. He had to move about the intersection, to light the walls. There were five tunnels, not four.

She quickly stepped back into the tunnel from which they had just emerged, to not lose it. Her mind raced. Somehow, with a real problem to deal with, she was far calmer than lying alone in the dark waiting. They could check all the tunnels, starting with the second to the left as the most likely. But they needed to make some kind of mark.

He lifted his mobile. "If only there was reception down here," he muttered.

She looked up. "Not down here." She pointed up. "What about up there?" It had risen meters, at least. How deep were they? They had gone down and up so many slopes, it was impossible to tell.

He followed her gaze. He lifted his phone, but the pale light barely showed anything. "I remember it getting narrow pretty fast. Good idea. I'll give it a try." He handed her the phone. "You hold this for the moment."

"Wait. We need to mark the tunnel we came out of." She did not want to get any more turned around than they already were. She backed further into the tunnel.

He picked up a rock. He joined her and scraped the rock along the left-hand side of the tunnel. It made pale scratches on the wall. She added a big arrow, just to make sure they would not mistake it for anything else.

He stepped forward and looked up. "Now, I'm going to have to get up there. I'll stand on your shoulders."

"Okay." She was not sure how that was going to work.

"Stand with your arms against the wall, like this, to brace yourself." He climbed on top of one of the large rocks leaning against a wall and put his arms out, like he was pushing on the wall. "I'll climb up."

Dubiously, she climbed up next to him and followed his instructions. He was lighter than she expected, if they swayed a bit.

"Now stand straight up."

She slowly straightened. She grabbed his lower legs, automatically, as if she could use them to brace herself. But it worked.

"Right," he said. "Give me the phone. I'm going to see if I can reach the far side with my legs and chimney my way up."

The phone was plastered against his left leg, in her hand. She carefully released her grip and raised her hand. He took the phone. His weight became sharply uneven, then suddenly he was gone. A moment later he came crashing to the ground below her.

"Damn." He stood up and brushed his seat off. "Again?"

They tried three more times. He stayed seated after the third. "It's so close. I can get my toes on, but just not enough pressure to brace myself."

"I'm taller. Maybe I could try," she said, surprising herself. She had no clue what she was doing. But she was much happier with the idea of climbing than being in tunnels. Although climbing in tunnels? She was not sure where that fit.

He stood up, and she lit him with his phone. "We'll give it a try." He took the phone from her. "Just the same as I did, crawl up my back." He leaned against the wall.

She grabbed his shoulders. "Um, what if you squat and stand up?" She did not think she could climb him like he had her.

He squatted, and she got her knees on his shoulders. He stood, slowly, and she balanced her way on the wall in front of them. "Now stand up," he said.

She shifted one leg, then the other. She wished the wall had handholds or something. But then she was standing in the darkness, one hand against each side of the stone wall.

"Here, take the phone."

She put her hand down. "I can't reach."

"Bend your knees a bit."

She did, feeling like she was going to topple off him. Then her hand met his, and she grabbed the phone. She stuck it in a pocket and everything went dark. She pulled it out again. "Um, what now?"

"What you want to do is get your back on one side and your feet over to the other wall."

She looked back and forth. "Could you, um, scoot closer to the wall?"

He did. She swayed. "Slower!" How was this even possible? But now she was nearly leaning on one side.

"Lift your leg off me and put it on the wall. I'll break your fall if it doesn't work."

She hadn't even done that for him. She lifted a leg. She tilted and jammed both legs out, jack-knifing in panic. But her feet hit the wall, and she stayed. "I did it!"

"Great!" She could hear real joy in his voice. She wondered if they were more than the 'little bit' lost he had claimed. "Now push your back up a bit, then shuffle your legs, and so on."

* * *

Carolyn's legs burned. How far up was she? Mike had vanished below in below in darkness. At least she had no gauge for how far she would fall if her legs gave way. It made her calmer.

"Still okay?" came from below. He would be alone there, in the dark.

"Yup." She looked at the phone. Still no signal. She scooted farther. The stone pressed, uneven lumps, into her back. She was sure to be bruised after this. She wished the wall was smooth; but then it would be harder to stay in place. She decided she was fine with the lumpiness.

A faint rumbling sounded above her. "Did you hear that?"

"What?"

It repeated. "That?"

"Didn't hear anything," he said.

The rumbling came again. She kept up her scooting, and the rumbling got more clear. There were multiple components: a low rocky, grumble, then a high-pitched rattle, and maybe some sort of smoother, whooshing sound. A strange shadow appeared in the tunnel above her.

Panic seized her; the rumbling was descending on her! Rational thought took over: the rumbling was no louder, and the shadow was static. "There's some kind of ledge." Yes, that was what it was. She played the phone around. The ledge was about forty centimetres thick and poked out from the wall across from her, fully filling about half the tunnel.

"Can you get on it?"

"I don't know." She kept scooting, until her knees were under the ledge and her body mostly above it. The phone light showed piles of shadowy items and graffiti on the walls. Would it hold her? She lifted the phone. One bar appeared. "I've got signal!" Her heart leapt. They were going to get out of this.

As long as the next thing she did was not fall to her death. She brought the phone back down, and the bar vanished. She would have it if she could stand on that ledge. "I'm gonna try for it." At least she was in this orientation. Had her head been on the other side, she would have had to rotate around first to get where she was now.

She flung her body forward. For one terrible, terrifying moment, she fell. Then her chest slammed into the ledge. She kicked back against the wall behind her and propelled herself fully onto the surface. She slid to a stop in a pile of buckets, ropes, and blankets. "I did it!" She stood up. Three of the five bars. "I've got signal!"

Her elation finally faded enough for her to realise she had not heard any reply from below. She kneeled down and leaned close to the edge of the ledge. "Mike? Can you hear me?"

"Barely," came what sounded like a whisper.

"I've got signal!" In its repetition, the phrase sounded senseless, like an advertising slogan. *What have you got? I've got signal!* She suppressed a bubble of hysterical laughter.

"Great!" said the whisper from below. The joy in Mike's voice was clear. He had been much more worried than he had let himself show.

"Who do I call?" Her only thought, dialling emergency services at 112, might not be the best plan. Not if the police were owned by the bad guys.

"Agent Dorsey is in my contacts. Try him first."

She stood again and flicked through the contacts. His phone had a lot of stored numbers. A small part of her mind noticed the ratio of male to female names was quite high. That observation made her feel strangely happy. Then 'Beane, Jenny ♥' flashed past, and her heart sank. She berated herself. What was she doing, thinking about Mike in any other way than a collaborator in this extreme situation? Surely she had far more to worry about than what he thought of her. And who Jenny-heart was.

Her self-recrimination was halted by reaching 'Dorsey, Adrian'. The small part of her mind that had noted the gender ratio was now distracted by Agent Dorsey actually having a first name. She tapped his name. It rang three times before Agent Dorsey said, cautiously, "Mike?"

"It's Carolyn," she said.

"Where's Mike?"

"Below. He's fine. We got a bit lost. I've climbed up some chimney thing to get signal."

"You're still in the catacombs?"

"Yes. Are you okay?" she asked, somewhat belatedly, she realised.

"We're fine. Okay, hold tight." There were some typing-like noises. "I've got a lock on you."

"Can you get us?" *Please,* her mind pleaded.

"Yes. It'll have to wait for nightfall, though. You're under a pretty busy thoroughfare."

She looked up as what now resolved to obvious sounds of traffic passed again. She squatted and leaned over to let Mike know the good news.

Chapter Twenty-Nine

Carolyn warmed her hands around her post-dinner cup of coffee at the cafe. She sat at the same back table as before, with Agent Dorsey, Mike, and Albert. It being earlier in the evening, the cafe was quiet and considerably more cafe-like, including her imagined flowers. Their table was centred with some kind of blue-tipped, trumpet-shaped blooms with a sweet scent.

Her brain slowly came back online, having mostly concentrated on eating—with all her terror in the tunnels, she had not realised just how hungry she was. Her last meal had been breakfast. The food sat happily in her stomach, and now her body was finally relaxing after the ordeal.

"They'll think we've got something," said Agent Dorsey. He and Albert had been as distressed as Carolyn and Mike when they shared the fact of the empty safe.

"How could it have been *empty*?" asked Mike again. "Just all that work with the message …"

Her mind finally made some new connections. "This message—when did they write it?"

"What do you mean?" asked Agent Dorsey.

"As in, is the message from when they left on their trip, or could it be from before? And they took whatever it was with them."

"I'm sure if they had it with them there wouldn't be all this trouble now." Mike tilted his head. "But good point." He squinted at Albert. "That was early stuff, wasn't it? As soon as they arrived?"

She took a sip of her coffee. "So they came from Bucharest with their secret, hid it in a safe, and left instructions to it."

"I think it pre-dated their arrival, in fact," said Albert.

She raised her eyebrows. "They sent their secret on ahead? Perhaps they did that again?"

"Again, how?" asked Mike.

"Perhaps they sent it ahead of them to Lihue."

"Lihue?" Mike sounded surprised. "In Hawaii?"

"That's where Ethan said they told him they were going …" She trailed off. Ethan had not told anyone else everything, she recalled. The Vandals, and probably Mike, Agent Dorsey, and Albert all thought the secret immediately preceded the move the Paris. Just as she had surmised from seeing the Paris lab books. What else had he not told them?

"Right," said Agent Dorsey. "He did say that."

She relaxed. She had not inadvertently revealed Ethan's reticence with information. She spun her coffee cup around. She was continuing to be reticent. But did it matter? Whether or not the secret was discovered before her birth, as Ethan said, or in the final days of Bucharest, did not matter for finding it. She found she was not jumping to share. Mike had gotten Ethan into a world of trouble. Albert had kidnapped Susan. Never mind how she and Mike had nearly been lost in the catacombs. She was still not convinced these people could be trusted.

Mike looked across at Agent Dorsey. "Lihue, then?"

"Lihue." Agent Dorsey lay a palm on the table. He looked at Albert. "Can you get us to the airport unseen?"

Albert flattened his hands on the table. "You'll have to enter surveillance in the airport proper, but we can get you there."

"I don't suppose I can convince you to pop on back to D.C., can I?" Mike asked Agent Dorsey.

"I'm not to let you out of my sight, remember," said Agent Dorsey.

Mike sighed. "So it will be three of us." He sounded resigned.

Three: Mike, Agent Dorsey, and … her? "Um, am I supposed to be coming?"

"You have anything better to be doing?" asked Mike with a smile. "If there's another safe, we'll need you. If there's a lab …"

"You'll need me to make use of the science." She finished the last of her coffee. "What about Ethan?"

"He'll be safe here for the time being," said Albert.

"I should not have involved him in the first place." Mike wrung his hands. "I'll fix things for him." This last seemed directed at Agent Dorsey.

"When do we leave?" She could use a rest. And to sit and have a chat with Susan. Susan would not have been terrified in the tunnels like she had been. Plus, she wanted another woman's perspective on her confused thoughts around Mike—thoughts she should not be having at all.

Albert stood and rubbed his hands together. "Now. If they think you've got something, they could be scouring feeds of departure ports."

"But Susan ... My stuff ..."

"Susan is not necessary for this trip," said Albert.

"She's my friend!"

"The fewer involved, the better," said Agent Dorsey.

Mike raised his eyebrows and flashed a smile at Carolyn. She could not help but return it, thinking on the two Americans' difference of opinion on Agent Dorsey's presence. But her amusement layered over unease. "I'm not sure ..."

Albert leaned onto the table. "Susan is settling well here. She knows Europe, not the US. She's better off here, and you're better off with those two."

She nodded soberly, feeling something like a toy being passed from one child to another. She had little choice, did she? "My stuff ..."

"We'll get you going, and someone will meet you on the way with your things." He looked behind him. "Let's move."

*   *   *

She pushed her knapsack under the seat in front of her. She wanted to poke through it to see what was inside; there had not been a moment since Nine or whoever that was handed it to her on the rushed trip underneath Paris. Someone curiously examining their own bag would trigger enhanced scrutiny of any surveillance video. That would still be true.

She leaned her head against the window. Mike sat next to her, and Agent Dorsey elsewhere in the plane. They were on an economy low-altitude flight. It would be another twelve hours or so before they landed in Los Angeles for the connecting space-hop to Lihue. Mike and Agent Dorsey had discussed the travel plans, but most of what she had come away from it with was that she had a chance to sleep.

She closed her eyes, barely able to believe she was now heading off halfway around the world with two American agents she had met only the day before.

Longing for her home and for Ellen suddenly overcame her. She clasped her hands tightly. She wanted everything to go back the way it was before.

Chapter Thirty

Carolyn exited the autocab and followed Mike and Agent Dorsey through a floral-bedecked path that merged into the interior of the hotel with no obvious demarcation of inside from out. The air was warm and imbued with a near-overwhelming honeyish scent of flowers. Mike peeled off and came back with keys. He handed one set to Agent Dorsey. "Adjoining rooms." He cleared his throat and faced her. "We can give you one room. As long as you keep the door open, for safety."

Their rooms were on the ground floor, along another inside/outside corridor that was separated from the pool area by a flowering shrub. Agent Dorsey entered one room, and Mike led her to the other. He opened the door and walked in first, leaned into the bathroom, then waved her in. The room held two double beds covered in blue and yellow floral bed dressings. Paintings showing ocean views hung on turquoise walls. Mike checked something in his palm quickly, then crossed to the connecting door and unlocked it. Agent Dorsey came in.

"Clear," said Mike.

"Clear," said Agent Dorsey.

She presumed this meant they were finally free of all surveillance, even video, and slung her knapsack around to the closest bed to investigate. She pulled out changes of clothes, the two mitocyls, a plastic bag full of toiletries, another bag with what looked like some sweets, a crumpled piece of real paper, and an elly-book. She flipped on the elly-book; it was the student's one. She supposed Susan had kept the one with the lab books. Or maybe Ethan had—he may have wanted something to occupy his mind while stuck underground in Paris. Reading a few year's worth of lab books could do that.

She uncrumpled the paper. It had a short note. *Sweets, K? Love, Suz.*

Mike peered over her shoulder, and she showed him the note. Albert must have passed messages back, and Susan packed her bags. Nice of her to add sweets, if a bit random. She sat on the bed. She missed Susan.

"What now?" she asked.

Mike and Agent Dorsey looked at each other. Agent Dorsey said, "I'll see what I can find out. The Bureau here should be clean, but I'm suspicious of everything nowadays. I'm on the Schwarz case, so looking after old Vivcor locations will be in keeping with that." He thrust his head towards her. "It's travelling with her that's not."

Mike nodded. "We'll keep a low profile."

Agent Dorsey left. Mike handed her the 3D remote. "Why don't you chill for a bit? I'll do some research in the other room and give you some peace. I imagine you want it."

"Can I visit the internet?" Signs had indicated the room was equipped with internet on the 3D.

He sucked on his lips. "Let's not. I'll check things out on my encrypted machines, but we don't want anyone curious about searches or visits from the hotel."

She touched the elly-book but said nothing. Neither Mike nor Agent Dorsey had shown evidence of being aware of elly-tunnels. Somehow she did not want to share that she knew a way to get to the internet unseen.

She pushed pillows up on the bed to make a backrest, sat down, and turned on the 3D. Mike retreated to other room. She flicked through channels, unsure where to stop. It seemed odd to be engaging in such a normal activity after the past week of hiding and running. She paused on the news. With her new perspective, reports regarding corporate activities took on a sinister tone. What was behind that new library, the current 'pubic interest' story? Was it really just beneficence of the company, or were they using the information of each visitor's searches in some way?

Was she being paranoid? Bae had expressed shock at Sandslin funding her. And Sandslin were the Vandals. Why hadn't that fellow—Nguyen?—just phoned her in the first place, instead of after the search of her lab? Or perhaps he only phoned because they had not found what they were looking for. She sat up straight, remembering him joking about the Schwarz Final Findings. That had not been a joke, had it?

She slumped. If she had internet, she would look up more about Sandslin. Mike was right: that would probably not be the best way to stay hidden, if anyone was monitoring the hotel's internet traffic. She rested her hand on the elly-book. What was the code for their other elly-book? She could contact it and tunnel. Perhaps that was why Susan had kept one behind: they would be able to chat, too.

She looked at the open door into Mike and Agent Dorsey's room. If Mike went out, she would give it a try. If she could remember that code.

*   *   *

Mike was still in the other room. Agent Dorsey had come and gone briefly. She picked up the elly-book and paged through the apps. There were some games. She opened one. It involved colour-matching to save some cute rabbits, and each level she passed built more of a hutch. It was strangely addictive.

Mike passed the door, apparently to look out his window, and back again. He did not seem to pay too much attention to her with the elly-book and the 3D on at the same time. She bit her lip. If only she knew the code. All she could remember was that it had a K as its only letter. Susan had sent her the elly-book; perhaps she had left the code in some way. She went back to the main page and examined the apps with a new eye. The Notes app was full of mostly course notes, but one file was hopefully called *Into the Tunnel*. The last edit on the file was months ago. Although could Susan modify such things with her police-tech knowledge? She opened it. It had five sets of four-character codes, all with names after them. None of the names were Susan, and none of the codes had a K.

She exited and looked further, but nothing else so obvious presented itself. Had Susan assumed Carolyn would remember?

Mike leaned through the doorway. "Are you hungry?"

She started guiltily. "Yes, actually." They had had lunch on the space plane, but that was hours ago now.

"Dorsey says he'll be quite late. I thought I could go out and bring something back for us—if you don't mind being alone?"

"As long as you do come back." She smiled at herself. Mike and Agent Dorsey would have no reason to abandon her here; they could have much more easily not brought her.

"Don't worry," he said. "See you soon." He vanished back into his room, then she heard his door open and close.

She sat up. Perhaps she should check her bag again; Susan may have left the code hidden somewhere. She pulled everything out of the bag and laid it on the bed. She lifted the clothes and shook each item. She put the toiletries in rows. She felt around the inside of the bag and peered into pockets.

Nothing.

She sat down with a sigh. At least there were sweets. She dumped them out and absently lined them up in rows of the same kind. There were three types, in different coloured wrappers: red, green, and blue. She unfolded Susan's note and read it again. She flipped it over.

Feeling somewhat silly, she took it into the bathroom and turned out the light. No, there was no hidden, glowing text, either.

But it, and the sweets, were the only oddity. *Suz.* Carolyn had never heard Susan refer to herself by the nickname. Nor had she heard anyone else. The note did not make a great deal of sense, either. Why would you ask if providing the sweets was okay? Surely the recipient had no way to say 'no' after hauling them halfway around the world.

It did not say 'okay', either. It said 'K'. Like the sole letter that Carolyn remembered from the tunnel code.

Excitement rushed through her. Her fingers tingled. It had to be something about the sweets. She picked them up, one at a time, but the wrappers had no words. She put them back into their neat rows. Then it hit her: three colours, to match the remaining three numbers of the code. There were six reds, three greens, and four blues.

That had to be it! Six, three and four. But in what order? *Suz.* A three-letter version of Susan's name. She dove for the note and flattened it on to the bed.

Sweets, K? Love, Suz.

That was it! Six letters in sweets, four in love, and three in Suz. She grabbed the elly-book and opened the tunnel app. She typed in '6K43'. The numbers swirled away and revealed the split screen.

Hiya! appeared in a text bubble on the top half. *Wondered when you would get on.*

Chapter Thirty-One

Susan? Carolyn typed. It could be Ethan.

That's me, the screen replied. *Glad you got through. Ethan was worried you'd eat the sweets before figuring out the message.*

Almost! replied Carolyn. *Did you know there's a file here with her friend's codes?*

No. A pause. *That would have been simpler, wouldn't it?*

Some. She smiled. It felt good to be chatting with Susan, even if only in text bubbles. *What's going on there? What did they tell you?*

Not much.

She waited, realising she did not know which question Susan had just answered.

Ethan's playing around with the genes on the cards again. Do you think there's a safe in Hawaii?

She shrugged in reply, then laughed at herself. *Don't know. Hope so.* Something in the other room creaked, maybe. She froze. But nothing followed. *Look, I'm not sure how long I'm alone for. Can you look up stuff on Sandslin for me? And Mike and Dorsey?*

Will do. Want to schedule another chat? Just luck I had this open this morning.

She bit her lip. When could she be sure of being alone? Maybe in the middle of the night? What was the time difference to Paris, anyway? *Night time here, maybe midnight? Can you figure out when that is?*

Will do.

A sharp bang came from the next room, and thumps of two sets of footsteps entered. *They're back. Talk to you tonight.* She closed the app.

She swung her feet to the edge of the bed and stood up. The 3D was still on, making it hard to hear what was going on next door, but it really did sound like two people. Mike had gone out because he said Dorsey would be back late. What if it wasn't them?

Sudden fear dried her mouth. She started slowly shovelling things into her knapsack, somehow thinking she could take it and run. But no, the opening door was obvious from that room to this; the reverse would be true as well.

"Whose room is this?" said a man. That was *not* Mike or Agent Dorsey.

Her hands shook. She had everything in the bag but the sweets. Where could she go? There was nowhere to hide in the bathroom. There was a sliding window to a porch, but it was at the far end of the room, past the connecting door. She had not yet tried to open it. What if it was loud?

"Looks like both of them dumped their stuff here. I'll check the other room." Another male voice.

She had to hide. The wardrobe? Under the bed? She dropped to the floor. Some hotel beds did not have an 'under'. Cloth draped all the way to the floor. She stuck her hand to the right, and the cloth gave way. It was just a valence.

Relief flooded her. She scooted sideways, sliding under the bed and pulling her knapsack with her. Stale dust tickled her nose. She breathed slowly, fighting a sneeze, until the sensation faded.

"It doesn't look like they've done more than turn on the 3D," said the second voice, loud and close. "Oh, candy!"

The bed creaked downwards. The 3D turned off. In the ensuing silence, she heard crinkling of a sweets wrapper. *Those are mine!* she thought furiously. She bit her lip to forestall hysterical laughter. This was not the time.

Another weight sunk down on the bed. "So we wait?"

"Guess so," said the candy-thief.

Another thump on the bed. One of them must have lain back. "Can I have one? Are they French?"

"How would I know?" asked the candy-thief. Another crinkly wrapper followed, and she fought a surge of irrational rage. Strangers lurking in wait for Mike and Dorsey were more of a problem than the fate of Susan's sweets.

The 3D turned back on, and there was shifting on the bed above her. She lay her head down. She could be in for a long wait.

*   *   *

"I—" Mike's voice cut off abruptly. His spy-senses could probably tell that people had been in.

The bed above Carolyn shook as the two men stood up. Frustratingly vague shuffling noises travelled around the room. It seemed to last forever. Then three people shouted, "Hands up!"

Mike said, "Hey, what are you doing here?"

"Why'd you run off with Dorsey?"

"I didn't run off with Dorsey. He *followed* me."

"Yeah, right."

She relaxed. They knew each other. Maybe it would be okay.

"He did. Look, you haven't explained why now *you've* followed me."

"You haven't been precisely forthcoming with your reports, Michael."

"I've been busy," Mike said.

"What did you find in Paris?"

"Nothing." That was almost true, pondered Carolyn. But he had found her. "Followed the message to a safe, but it was unlocked and empty."

"Really?" Sarcasm seemed to drip from the word.

"Yes, really," Mike sounded annoyed.

"I find that hard to believe. Why did you hare off to Lihue, of all places? Right after opening the safe?"

"Professor Boltzer had told Dorsey that the Schwarzes were planning to come here after their final holiday. When we found the safe empty, it was the next place to check." Mike's voice took on a pedantic, slightly patronising tone. She was distracted enough by the change that it took her a few moments to realise that Mike had distinctly not said anything about her. Or even about talking to Ethan recently.

"Oh," said one of the interlopers. "You should have said."

"I did."

"No you …" The voice trailed off, sounding unsure.

"You've not read my reports," said Mike definitively. "Does anybody know you're here? What exactly is going on?"

"This is bigger than you know, Michael."

Another weight joined the bed. Mike must have sat down. "Enlighten me."

Two throats cleared. "Where's your friend?"

"Out doing whatever he does with his Bureau buddies, I presume. Why?"

"Perhaps it would just better if you headed back to D.C. We'll take over."

"You haven't explained a thing yet. I have no reason to change what I'm doing. I'm not even sure whose orders you're following."

More throat clearing. "You don't need to know."

"Then I don't need you stepping on my toes. Get out."

"You've got two rooms. You and Dorsey can share. We'll just take this one." *No!* Carolyn wanted them out, too. And definitely not to be stuck under the bed any longer.

"This is my room, and I am asking you to leave." Mike sounded menacing to her. She wondered if he made any impression on his not-friends.

"You don't know what you're doing, Hafal."

"Tell me."

The bed creaked. Were they standing up? "I really recommend ..."

"Out!" shouted Mike.

Footsteps clumped away, followed by the distinct loud clunk of the door opening and closing. Relief flooded her. She lay her head on the back of her hands in sudden exhaustion.

Several moments passed in silence. "Carolyn?" Mike's voice was small and unsure.

She jerked up and banged her head on the underside of the bed. "Ow." She scooted to the edge of the bed. "I'm here."

He lifted the valence and offered a hand to her. "Thank goodness."

She scooted out, then took his hand and let him guide her to sit on the bed. Her eyes lit on the candy wrappers strewn about the bed. Somehow the loss of Susan's sweets seemed the ultimate betrayal. "What ..." Her voice shook, and she could not get out the rest of her question. She felt her eyes filling with tears. Her face heated. Just like in the catacombs, she was going to weep when a danger was *passed*. She was embarrassed, and somehow, the thought of Mike thinking her a coward was highly disturbing.

He put an arm around her and gave a quick squeeze, then patted her shoulder. "It's okay. They're gone, and they didn't leave any devices."

She took several deep breaths, successfully fighting away the tears. She found herself wishing he would put his arm back. "Who were they?"

"Some of my colleagues. I'm not sure what's going on, but I think we're going to have to be more cautious."

"I noticed you did not mention me or Ethan."

"No." He clasped his hands. "I really regret bringing Ethan into this. I don't want to regret bringing you in."

"You didn't. Remember, Sandslin had me pretend-killed first." Her mind wandered to Ellen, and how Ethan reported Sandslin were completely ignorant of her. "It seems there's a lot of things going on here." She lifted her face towards Mike. "Would they have real-killed me if I showed up in London? I don't understand why they did that."

"I don't either," he said. "I don't know what's going on back home. I don't like not knowing."

She crushed an empty candy wrapper in her fingers. "They ate Susan's sweets." She missed Susan. She missed Ellen. She missed her life. It all came crashing down like a waterfall of sadness. Tears leaked out.

He put his arm around her again. "We'll get you more sweets."

She sniffed and smiled. "It's not the sweets."

"I know." He squeezed her shoulder.

She leaned into him, relaxing into the offered comfort. Then she remembered his phone and a name: Jenny, was it? She straightened and stood. "So, did you bring supper?"

Chapter Thirty-Two

Carolyn went to bed before Dorsey returned. She thought she would find it hard to sleep, but dropped off as soon as she pulled up the covers. The next thing she knew she was lying in a dark room with buzzing next to her ear.

The elly-book! She blinked sleep out of her eyes and swiped it on. She had set the alarm for a few minutes before midnight. It took all that time to fumble her sleep-addled brain to the tunnel app, the room lit by the glow of the elly-book. Belatedly, she pulled it under the covers and made a tent with the duvet, like a child reading illicitly at night.

Almost immediately upon reaching the cloud screen, the app displayed, *Bridge Lb31 Requested. Accept? Deny?*

Carolyn stared at the message, her brain like molasses. There was no K. That wasn't Susan. A spike of fear brought her to full alertness. Someone unknown was trying to communicate with her. Who could it be? Should she hit 'Deny'? She didn't know if that sent some kind of message back. Her agents didn't seem to know about the tunnel, but perhaps others did. She didn't want to reveal herself anymore than she needed to.

She quit the app, waited a few minutes, and restarted. This time the cloud screen popped up *Bridge 6K43 requested. Accept? Deny?*

She hit accept with a sense of relief. *Hiya! How was today?*

Scary! replied Carolyn. *Some of Mike's colleagues showed up. But he didn't tell them about any of us. He seems worried. I am too. And someone else tried to tunnel me.*

Hmm. Interesting. I wouldn't answer any calls except mine. Do you remember the code?

No. Her face flushed. She should have made note of it. Maybe Susan could have found out more with that. *I won't answer.*

The screen was still for what seemed like a long time, but was probably no more than twenty or thirty seconds. Then: *Okay. I've got some stuff for you. I'm going to copy and paste it in, and I think you can grab it out the same way. Don't try to read it now. Ready?*

She experimentally scrolled the screen, but their conversation did not go off the edge yet, so she could not tell how that would work. She tried selecting Susan's last message, and it seemed to copy. She flipped over to the Notes app and pasted—yes, it worked.

Ready, she typed, back in the tunnel.

The screen filled with text, and she spent the next several minutes simply selecting, copying, and pasting. *There's some light reading for you :)*

:), she typed back, finding the old-style emoticon strangely nostalgic.

You should get some sleep, typed Susan. *Same time tomorrow?*

* * *

She woke confused, with something hard on her cheek and weight on her head. She flailed about and sat up. Memory rushed in, reminding her she was in a hotel on Kauai, in hiding from Sandslin. Mike was hiding her presence from his co-workers or whoever they were. She had spent the middle of the night chatting with Susan, hiding that fact from her companions. She rubbed her eyes, wondering if she would be able to keep straight who knew what.

She itched to read the material Susan had found for her, but she could hear Mike and Dorsey moving about in the next room. She would need privacy.

Mike leaned into the room from the door. "You up?" He was looking at her, so it was an obviously rhetorical question. "The hotel has breakfast, but I'm not sure you should be advertising your presence here. Dorsey is going to bring something back for us, if they let him. Go ahead and shower or whatever." He cleared his throat. "We do need to the keep the door open, for safety, but I'll stay around the corner. Let me know when you're all up and dressed."

She nodded, not quite feeling ready for speech. He vanished back into his room. Their door clunked open and closed: that would be Dorsey, going for breakfast. She rested her hand on the elly-book. Mike said he would give her privacy. But he might check if he didn't hear the shower and everything, in case she had fallen back asleep.

She stood and made her way to the bathroom. She paused in admiration of the shower, not having truly looked before. After the spare, communal rooms in Paris and the chilly bathhouse in Bucharest, even her own flat's shower would have been a luxury. But the Hawaiian hotel had a shiny chrome shower-head with five settings, including massage; plus, those jets around the edge of the sparkling white tub with their touch-panel controls looked very enticing. She turned on the shower and stepped into the warm spray. Maybe she could have a bath tonight.

She left the water running when she reluctantly got out. She grabbed the elly-book and sat wrapped in the hotel's thick robe on the cool tile floor. Steam wafted past, and she leaned into the bath to turn the shower tempera-ture down. She did not want to damage the elly-book. She resettled herself and opened the notes from Susan.

Dorsey appeared to be a legitimate FBI agent. He had posting info in the main office in Washington, D.C., if little else. Mike was a ghost: Susan left only a short note to the effect that he did not seem to exist at all. Carolyn sup-posed that was to be expected for a spy.

Then there was Sandslin. Carolyn felt queasy reading the material. She had taken money from these people. But that was for basic research! Not for what-ever strange goings-on were occurring in less-developed countries, where Sandslin appeared to have swallowed entire villages. Youngsters in the West for education had returned home to find families relocated or vanished, par-ents suddenly unwilling to discuss their jobs, and a strange dearth of feral dogs. It felt like the set-up for a bad science-fiction film. Beyond the weird stuff, there was the typical corporate greed. Working conditions only at, or sometimes just below, mandated standards. Several waste disposal scandals, with fines later waived on technicalities.

She wrung her hands. She had justified a lot for the progress of science and her own small career. She was aware of the anticorporate movement, how could she not be? But it seemed a fringe activity, radical and hysterical. Yet the families in the tunnels in Paris did not seem hysterical. A bit radical and fringe, though. She felt her world shifting again, as it had when she learned the police and the criminals were working together, and later when finding governments and corporations were at odds.

The information age of her youth had passed into a propaganda age run by corporations, yet most of the populace was blind to this. As a scientist, she knew how much of the 'open access' work was owned by corporations and how quickly it could be removed from public view. Yet they did not remove it—corporations needed the good view of the public to continue working behind the scenes, collecting and owning everything.

It had seemed okay, before, to rely on the beneficence of corporations. But that was before she had seen this side of them. She had been so grateful to Vivcor, for helping her change her name and hide her infamous identity. Those days resurfaced in memory, how she had been so concerned to get everything wiped before she started University: so she could start on a clean slate, as a new person, without having to answer questions about being the 'Human Hoax' to all her new University friends. She had been so concerned that it would happen, she had not really thought on how *easily* it had happened. All her records, all her social media, everything begun anew. Even back then, corporations had held the information, hadn't they?

She felt chill, despite the warmth emanating from the running shower. It had not been purely altruistic of Vivcor to erase her identity. Letting her fade away let their embarrassment fade away. The first real synthesized humans had not been far away, and perhaps Vivcor thought they would be the holders of the true thing.

Her breath caught. Surely they knew. Surely they had known all along. Think of all the designer babies, all the actual, real use that had been made when synthetic human technology truly appeared. Vivcor had been unable to capitalise on her, because she wasn't real. That was not something they would have been able to ignore. They had let her remain in the dark, acting as ambassador and biological curiosity, showcasing Vivcor's amazing (not-)success.

They had been able to, because no one asked the obvious question of what use is this advance? Or no one had asked and the question allowed to be heard.

She wondered if, somehow, things were restored, could she go on with her life and research as she had before? Perhaps not. Perhaps she did not just want her life back. Perhaps she wanted something different.

Chapter Thirty-Three

Her stomach was rumbling by the time Dorsey arrived with breakfast. She eagerly joined the agents in their room, sitting at their small round table near the curtained sliding windows. Music played unobtrusively from somewhere near the door. Dorsey pulled a variety of fruits, some croissants, and a small pile of sliced cheeses from a sack and spread them on the table.

"What took so long?" asked Mike.

"I got a bit distracted," said Dorsey. "I didn't come straight back after breakfast, sorry—but I've got info. My Bureau buddies handed across something that could be a lead about where to go next."

She fingered an unfamiliar spikey red fruit curiously, then made a sandwich with the cheese and a croissant. "Which is?"

Dorsey began, looking at Mike, "Your friends ..."

"They're not my friends," said Mike, reminding Carolyn of Ethan's protest when she named Sandslin similarly.

"Your colleagues," said Dorsey, "arrived shortly before us. They probably took a space hop from Washington when we got to LA. Bureau teams tailed them as soon as they arrived. Something's up with Vivcor and Sandslin and Maxtech. Your fr—"

"Who's Maxtech?" interrupted Carolyn.

"A big twenty subsid. Not a friend of Sandslin: long-time competitors. Potential connections with Vivcor, at least a decade or so ago."

"Go on." She had thought there might have been another player in this. The people who had shown up at Uncle Keith's could have been Sandslin, and she supposed she have moved towards assuming they were. But they might not have been. Finished with her sandwich, she lifted the red fruit again. "How do you eat this?"

Dorsey took one and jabbed a fingernail into the skin, then tugged, peeling it. The inside was a translucent white. "Take the skin off like this. You eat this bit—but be careful, don't eat the seed." He took a bite.

"My colleagues were?" prodded Mike.

"Your colleagues have been searching through old Vivcor sites. Maxtech is either following them or working with them ..."

"My colleagues wouldn't ..." Mike trailed off, his face contemplative. Perhaps he was rethinking what they would or wouldn't do. She freed the inside of her fruit and nibbled experimentally at the white flesh. It was sweet, sort of like a grape crossed with a pear.

"In the meantime, Sandslin has arrived, but they seem more in the dark. They've not crossed paths with your friends or Maxtech, although I have difficulty believing they are truly unaware of them. They've been concentrating on a currently active Vivcor site, but sneaking about like they did in Bucharest."

How did Dorsey know what Vivcor had done in Bucharest, she wondered. She tried to remember what she, Ethan, and Susan had told them. Yet his words almost suggested it was something he had observed for himself. She gave herself a mental shiver. She was starting to be suspicious of everything and everyone.

"Okay, so how does this tell us where to go?" she asked.

Dorsey cleared his throat. "It doesn't exactly. But I was hoping Mike could work his magic again." He pulled a rolled docfilm out of his inside jacket pocket and flattened it on the table. "This was at the old site. It looks similar to the note you followed in Paris."

Mike put aside a banana peel and leaned over the docfilm. It was a photo of a piece of paper, filled with handwritten numbers. They came in pairs, separated by commas, with dashes between the pairs.

She squinted. She recognised those number formations from recipes in her father's lab books. Excitement pulsed through her. "That's my father's handwriting!"

Mike pulled the docfilm closer to himself. "Yes, it looks like a similar code." He sighed. "Is this all? It's a lot shorter than the last one."

She pulled her mother's ID card out from inside her shirt where she had been wearing it. Those protein sequences had been handwritten; she had not paid attention to the handwriting before. She wondered if she could tell if her father or mother had written it.

"That's what your colleagues took away from a Vivcor site just last night," Dorsey said.

"Damn." Mike made a fist. "They'll have decoded it by now." He clasped his hands in front of his chin and leaned on them. "Perhaps. I'm not sure they've actually read my reports." He unclasped his hands to cover the lower half of his face, then dropped them. "I'm going to have to call it up again, to access the code." He looked at Carolyn looking at the ID card. "What's that?"

She held it out towards him. "My mother's old ID for Vivcor Paris. Remember?" She had passed on the information she had gleaned about the protein sequences on it and her father's card. "But it looks like …" She glanced at the note on the docfilm, and back to the card. Yes, the handwriting was the same. "… it was my father who wrote the protein sequences on the back."

"Damn!" Mike said again, but with a distinctly different emphasis. He held his hand out. "How did I not … Can I see?"

She threaded the chain over her head and handed the card across. He scanned it, running his finger alternately along the lines and rows, flicking his eyes to the docfilm occasionally.

"Well, I'll be." He laughed. "I don't need to find my notes after all. This is the code."

"How?" she asked.

"There was another one of these, right? Exactly the same?"

"Almost," she said. "This one is in reverse—starting from the bottom right. My father's started from the top."

"So they made some attempt at obfuscation, at least."

"How's it the code?" she asked again.

Mike placed the card on the table. "It's a simple substitution code over plain text, based on a degenerate key: this, apparently. The first number is the row of text, the second is the number of characters in. Then you just read the letter off. The numbers fill in those letters that seem to be missing, like O and U …" His voice rose in a question, as if he were wondering why they were missing.

"O and U don't mean any amino acid, so they wouldn't be there," she explained.

Mike tapped the card. "It makes words."

She shifted to squat on her knees on the chair, leaning farther over so she could see the docfilm better. "What does it say?"

"Hold on, this is a bit slow!" Mike's voice held laughter. She was unsure if it was at her, or just in joy at being able to decode the message so easily.

He slid another docfilm next to Dorsey's and wrote, first a quick burst of letters, then slowly consulting the coded note and the card alternately.

FOLLOWTHEMYNAHBIRDTOPUFFSHOMELOOKUNDERTHESADROOM

She read: "Follow them Y-nab irdt … Follow the mynah bird!"

"To Puff's home," continued Mike.

"Look under these … the sad room," finished Carolyn. She grinned. It was a message from her parents. Somehow, it seemed closer and more real than all the lab books she had read. Those were just records. This … this was a message. Intentional communication.

"I see why it's shorter." Mike looked glum.

"Is it similar to the last one?" she asked.

"The last one had very detailed directions; we just had to follow them. This …" He shrugged. "It does look like directions, but I have no idea what it means."

"Perhaps they got better at codes," she said. Mike smiled; she felt good to have made that happen.

Dorsey leaned in. "Well, it seems like some kind of references. Mynah bird? Puff's home? Sad room? They must mean something."

"To each other." Mike looked at Carolyn. "Anything in those lab books of yours that would help?"

She shook her head, but her brain supplied something else: a tune, a soft female voice. "Turn that music off!" she said. Dorsey jumped to comply. She blinked at him, mildly disconcerted both by her temerity and his rapid response to her command. But the tune repeated in her mind, and her chest felt warm, comfortable: a rare, early memory, of her actual parents. She had so few. Tears wet the corner of her eyes. She closed her eyes and began singing along with the voice in her head: *Puff the magic dragon* …

Minutes later, she opened her eyes. Mike was scribbling on his docfilm, and Dorsey staring at them both in silence.

Mike stopped writing. "'By the sea' doesn't particularly narrow it down on an island."

"Well, it does some." She still felt dreamy from her memory. She wanted to replay it, over and over. Such a beautiful, sad song: more so being sung in her memory by her dead mother. "He went to a cave when he was sad. Are there caves along the shore?"

"Yes!" said Mike, excitedly. His face fell. "All around the island."

"Oh," she said.

Dorsey turned and spoke into a device into his hand. He retreated to far the side the room. Carolyn and Mike watched him. Finally, he dropped his hand and crossed back to them.

"What is it?" asked Mike.

"Your friends," said Dorsey—Mike grimaced but said nothing, "have just left for Maui."

"Maui? Why?"

"They were saying something about a Willy Kay and Dragon's Teeth. Mean anything?"

Mike pulled over his computer and typed furiously. "They've clearly decoded the message. Dragon's Teeth matches with Carolyn's song here—it's a formation along the north shore of Maui—but I don't know about Willy Kay. Ahah!"

"What?" She leaned over, trying to look at his screen.

"Willie K—just the letter—was a musician at the turn of the twenty-first century, from Hawaii. He came from Maui … and … had a song called 'North Shore Reggae Blues'. Look here." He gestured at his screen, which was showing lyrics. "The mynah bird leads the singer to the north shore."

"So we go to Maui," said Dorsey. Mike nodded.

Excitement flushed through Carolyn. They were on the trail! But it felt wrong. "I'm not sure," she said, hesitantly.

Dorsey scowled, but Mike turned to her, face curious. "Why not?"

"Well ..." She gathered her thoughts. "Yes, Puff was a dragon. But it says 'Puff's home', not 'Puff's teeth'. And ..." She frowned, trying to capture the logic behind her sense of unease. "I don't think it would be on another island. If this is a note to a safe anything like that last one, it wasn't a small item. I imagine it could have been shipped here with their lab stuff, but if they then hid it, they would have had to move it themselves. If they were keeping it a secret, transporting it on a boat or a plane wouldn't be particularly inconspicuous." That was it. "I think it's on this island. The mynah bird leads to the north shore. Any caves there?"

"Yes," said Mike. "Three: two wet and one dry. I think ..." He typed on his computer. His eyes lit up. "Yes! One of the wet ones is famous for something known as the Blue Room!"

"The sad room," she said, nodding.

Dorsey retreated to the far side of the room again, hand to his mouth. He returned. "Maxtech and Sandslin have just gone to Maui too." He scowled at Carolyn. "Are you sure about this?"

She stared at him. Of course she wasn't sure. She had no idea what she was doing at all. But she tried to imagine her parents: one of them, probably, come to the site ahead of time while the other stayed behind with their child—her. Tears welled again. She had spent so much time being angry at her parents, first for saddling her with celebrity, later infamy, then, as her own career progressed, for being dishonest scientists. She had not thought about them as people so much—as parents. Maybe there was some reason for their deception; maybe related to this secret Ethan said they'd found, that might be sitting right here, in a safe—or something—not far away. But her mother or father, alone on the island, or with a few trusted lab members: they would be hiding whatever it was in the most inconspicuous way possible. "It's got to be on this island."

"Okay." Dorsey crossed his arms. "I think we should go."

"Now?" she asked.

"Now," agreed Mike. "While everyone else has left. Did they leave anyone behind?"

"Not that the Bureau can tell. Everyone they've been tailing took the flights."

"Okay, we head out." Mike stood. "What have you told your buddies about Carolyn?"

Dorsey looked at his hands. His face grew shiny, with a slight pinkish hue. "I said she was one your colleagues."

Mike raised his eyebrows. "So you don't trust them, then."

Dorsey looked straight across to Mike. "I'm not sure I trust anyone not in this room."

She swallowed. Mike did not trust his co-workers, nor Dorsey his. But they trusted her. And here she was, keeping secrets from them. But it was not the time to burst out that she had secretly communicated with Susan. The material Susan sent her was not anything either of them would have trouble getting: just their own (non-)histories and publicly available information about Sandslin. It was not like their trust in her was misplaced.

Chapter Thirty-Four

They parked the hired car in a lot along the shore. Carolyn stepped towards the sea. The sound of the surf reminded her of walks with her uncle along the Fife coast—such a different shore, yet somehow having a similar quality. She hugged herself, trying to figure out what it was: not the turquoise blue water, or the palm trees, or the sun-blushed warmth. A breeze tugged at her. Something to do with the wind: the wind and a sense of freedom. That was it—she associated the shore with freedom. The expansive horizon: a limitless place to go. How limitless was it, really? With corporations in control of so much of the world, even here, in paradise. She thought about her parent walking along this same shore, some three decades ago. Had he or she had the same sense of freedom? Or was there inklings of the upcoming … murder? Not freedom, but a small box of fear, during which her mother or father hid something Carolyn hoped would free her now.

She shouldered her knapsack—she had stuck some of the hotel's bottles of water and the student's elly-book inside, loathe to leave her only connection to the outside world behind—and turned to follow Mike along a small jetty where a rank of water taxis parked. For now, they seemed alone and free of surveillance, other than perhaps Dorsey's compatriots. She squinted at the building on stilts to which the taxis provided transport. "Why's it way out there?" Mike had spoken of an aquarium where they could get an early lunch before hiking up to the cave.

"They have tanks connected to the open ocean," Mike said. "Apparently it was easier to build the aquarium around the reef, instead of bringing a reef to the aquarium."

The water taxi ride was short, and the fine spray refreshing in the heat. She stepped into the aquarium's wide entryway as Dorsey paid the taxi driver. What corporation owns this, she found herself wondering. She had never had thoughts like that before. Things just were: libraries, railroads, police. But someone owned all of it. And the owners could be connected in deep, disturbing ways. Like how the campus police appeared to be working with Sandslin.

Dorsey and Mike joined her, and they entered. A curved staircase framed an open space, with large signs: *Vivaculture*, at the top of the stairs, and *Aquaculture*, at the bottom. Aquaculture would be aquariums. "What's vivaculture?" But as the words left her mouth, she knew: "Oh, vivariums. Indoor gardens." Or was it whole ecosystems? Or was that terrariums?

"This used to be a Centre for Excellence on Aquaculture and Vivaculture, back when Hawaii was really big on bringing the farming into artificial environments," Mike said.

"But now it's just another tourist trap," said Dorsey.

"Not entirely. They still do some research here, I think." Mike looked at Carolyn, as if she would know.

She shrugged. She studied mitochondria, not plants and fish. But she wondered, if they did do research, whose research? Was it associated with a University? Was it corporate? Was it both, like her research: University work, funded by the only source that cared about basic research anymore? She hugged herself tighter as she followed Mike and Dorsey up the stairs. Why *did* corporations fund basic research? Because they had the long view, she had thought before. They knew that what was basic science today might be the restriction enzymes or green fluorescent protein of tomorrow.

Discoveries of basic biology had led to molecular tools that expanded the entire capability of biotechnology. Restriction enzymes were enzymes that bacteria used to protect themselves from viruses, cutting up viral DNA into little bits. But scientists used that ability to cut up DNA and combine together pieces from different organisms to do whatever they wanted. Green fluorescent protein came from a bioluminescent jellyfish, but was now used in a variety of slightly modified forms by scientists to tag anything they wanted to view under a fluorescent microscope. If someone hadn't asked, *How do these bacteria survive viral infection?* or, *How does this jellyfish glow?* no one would be using those now.

But was that the only answer? Did corporations really have the long view that governments, in thrall to their tax-payers, lacked? Or was it something else? Was it part of the creep of the corporations to own all the data and knowledge in the world? She thought about the communities living underground in Paris or in the unlikely wilderness in the centre of Bucharest. Maybe they had the right idea.

*   *   *

She waited with Mike at a table while Dorsey went to get everyone food. Suddenly, she felt awkward. Her brain spun through her interactions with Mike, too many consisting of her in tears and him consoling her. Why did her heart keep thumping like that when he smiled at her? *Why* was he smiling at her? She forced herself to reflect on that heart-tagged name in his phonebook. "What's so funny?" she asked, aggressively.

He shook his head, still smiling. "Not funny. Just …" He looked off into the distance. "I've been following the trail of the Schwarzes, and suddenly I find myself here with their daughter. A real, blood relation. I never really thought about them having a daughter."

"Um," she said, "them having a daughter was kind of the whole deal about their fame and then …" She trailed off. Their fame and then her embarrassment.

"I know." He folded his hands under his chin, but kept his smile. "I'm not getting my words right. Like before you were just an idea, and now you're a real person. With all the complexities of a real person."

She almost understood. It was similar to her shift in view of her parents, who had been, as Mike had said, ideas: the famous scientists, then the infamous deceivers. Thinking of them with toddler her, leaving each other coded messages, and in flight for their lives, was very different. Mike was still smiling, and she found the conversation a little too deep for her liking.

She looked around for inspiration. The cafeteria was situated on a platform behind the staircases, halfway up towards the vivariums. Below to her right was a glass wall that looked over a sunlit reef, and above greenery reached upwards into open air. Below to her left an aquarium showed a shipwreck scene, with fish flitting around spars, masts, chests, and various other objects. Above, the vivarium was bathed in fuchsia light. "Why is it pink up there?"

Mike's smile remained, but he turned his head to look upwards. "Growth lights. Looks more purple to me."

"Fuchsia," she allowed. "But why?"

"It's more energy efficient," he said. "They emit only those wavelengths that the plants actually use for photosynthesis. While we think of plants using 'sunlight', chlorophyll only absorbs a set of wavelengths in the blue, then again in the red. So together they look fuchsia."

"Ah," she said. "Sort of like how humans are sensitive to the blue in sunlight—they're always talking about blue light from screens messing up your sleep, and SAD lights are usually just blue."

"Sad lights?" He wore a quizzical expression.

"Not unhappy lights!" She laughed. "Seasonal Affective Disorder. When sunlight is limited in the winter, some people can get depressed. They can be treated by making sure to get light stimulation. It's fairly common in northern places, like Scotland, where I grew up."

"But they only need limited frequencies, like the plants," he said.

"Yes," she agreed. "Hey, I thought you said all you knew about biology was codons. Where'd this stuff about chlorophyll come from?"

"I garden quite a bit."

"That's definitely biology!" she said. Dorsey set down a tray with steaming mugs. Carolyn looked up with a smile, suddenly more relaxed. "It smells great. What is it?"

"Saimin." Mike took one mug. "It's a Hawaiian speciality. I always love getting this here."

She took one herself. It was a noodle soup, with slices of cooked egg and a faintly ginger scent wafting out. She tasted the soup; it was salty, with a mild taste of chicken and prawns. Not quite like anything she had had before. She liked it. "So, what's the plan?"

"We head out to the end of the road." Dorsey pulled a thin piece of docfilm from his pocket; it looked to be an advertising flyer. "Across from the parking lot is a path up into woods—the cave is there."

She pulled the flyer across to her. It showed a green-fringed cave entrance with a hint of glowing blue water inside; she could not tell if it was a real photo or an artist's impression. "Under the blue room? Or sad room, I suppose they said. Is there any under?"

Dorsey shrugged. "I think we need to go and take a look."

She frowned, looking at the picture. It seemed such a strange place to hide something; she supposed the underground junkyard in Paris was strange too, but at least it made sense as a place to hide things. Could her parents have possibly left something in a tourist destination—a natural attraction, at that—that remained undetected for three decades? Something did not feel quite right. Perhaps the corporations had had the correct idea haring off to Maui. Well, they were here now. They could look.

*   *   *

She drank the last bits of broth from her cup of saimin and put it back on the tray with Dorsey's and Mike's empties. She felt nervous, different than in Paris, more excited. It seemed relatively safer: they knew the Vandals—Sandslin—and everyone else were elsewhere. So now it was just on the trail of the mystery.

"Let's go." Mike led the way from the cafeteria while Dorsey cleared their tray. Mike paused at the base of the stairs, waiting for Dorsey. A small group of people were setting up a table along the left-hand wall. Anticorporate, Carolyn immediately characterised them, although she was unsure why. Perhaps it was a slight homemade feel to their clothes, but more the pile of flyers they were each lifting. Someone began unfurling a sign and tying it to the railing of the stairs: *Save the* ..., the first part of it said.

"Keep out the corporations!" A woman thrust a flyer towards Carolyn. Carolyn took it. It was real paper, not docfilm. That made sense: they wouldn't share anything with the cloud. She smiled at the woman. They were on the same side, now.

*   *   *

They piled out of the hired car. She tingled with excitement and nerves. They crossed the street and took a wide path that went slightly uphill into the jungle. The tropical idyll of the place seemed dissonant—surely this harrowing adventure did not have its conclusion in such beauty. The Paris catacombs seemed more fitting.

In barely two minutes, they stood above a large cave: it looked like a wide, partially-open mouth, with vines hanging down from the cliff above. No one had spoken in the short walk. They remained silent as they clambered down the steep slope into the cave. She looked behind her; the sky above the opening now looked a giant blue eye. Through the mouth into the eye.

They came to the shore of a wide lake. The water seemed to glow slightly blue. "*Under* the blue room?" said Dorsey.

Mike sat down and stared at the water. It was clear enough to see the bottom, which consisted of uneven rock. "Maybe back in shaded a corner?" He pointed towards the edges of the cave. "It must come from some underground source."

She sat next to him. "This doesn't seem right." She drew up the mental image of her lone, frightened parent again. "That thing in Paris was huge and heavy. The path up here was pretty smooth, but even so, they would have had to walk. It's not like it could be driven here. Carrying it would be difficult."

"It might not be the same kind of thing," said Dorsey.

"They did obfuscate their message more this time," said Mike.

"But not that much," said Carolyn. "It still used the same code."

"True." Mike gestured at the water. "This is where it got us."

"Perhaps we should have gone to Maui." Dorsey voiced Carolyn's thought from earlier.

But, no. She said, "Maui would have the same problem. Some kind of rock formation along the shore? In *Hawaii?* A hiding location that would have to handle surf, and storms, and hurricanes? We're talking a pair of geneticists here. Not secret agents."

"Or spies," said Mike with a grin.

She grinned back, somehow cheered by his joking even in all of this. She stared at the water. "Could we dive under or something?"

"We'd need proper equipment. And dry suits," said Dorsey. "There are infectious bacteria in these waters."

"We'd need to be respectful," said Mike. "This is a sacred place to the indigenous population."

Her parents wouldn't hide something in a sacred site, she felt. Although she wasn't sure why she thought that: they seemed fine to deceive the entire scientific community for their own gain. But … her thoughts from the night before came back. What kind of gain had it been? It clearly had not gotten them, nor Vivcor, riches. Just how many people knew about the deception, she wondered again. Why had they done it? Was it somehow connected to these Final Findings? Ethan had said the Findings were from just *before* she had been—supposedly—created. What could they have found, that would make them distract the whole world with a false tale of the first synthetically engineered human?

If it was something they were hiding from their employers—yet Vivcor must have known of the deception as she had surmised last night—what was the point? Factions within Vivcor: hiding the Final Findings from some, hiding the truth of her biological origin from others? What had it been like to work for a corporation in those days? Days when the corporation stranglehold on the world had been in the making? When corporations seemed more like individual entities, although conglomerates and subsids were well on their way? Before organised anticorporate movements?

Dorsey joined her and Mike in sitting and staring at the water. She pulled out the paper flyer she had taken from the anticorporate woman outside the cafeteria. She had shoved it into her pocket without reading. She flattened it on her knees.

Save the C.A.V.E.! it said.

She squinted and read it again. Cave? Excitement pulsed. She smoothed the rest of the flyer on her knee, hands trembling. It continued: *The Centre for Aquaculture and Vivaculture Excellence has been a public venture for over half a century. Yet now the Hawaiian government is considering …*

Half a century: fifty years. That place had been in existence when her lone, worried parent walked these shores. She leapt to her feet. "Not this cave!" She gestured about her. "This C.A.V.E.!" She flapped the flyer in front of Mike.

Chapter Thirty-Five

"What cave?" Mike slowly stood.

"The Centre for Aquaculture and Vivaculture Excellence!" Carolyn was unable to keep her voice from a shout of joy. "C, A, V, E: cave. We were just there."

"But the blue room," said Dorsey.

"The *sad* room." A vision of the fuchsia lights in the vivarium flashed in her memory. She grabbed Mike's shoulders and spun him to face her. "Remember how the growth lights made me think of Seasonal Affective Disorder lights—SAD lights? *That's* the sad room."

"I don't know …" started Mike.

"My mother would be familiar with them. She grew up in Scotland!" She scrambled up the slope. "Come on, this has got to be it!" She waited, impatient, at the top as Mike and Dorsey climbed slowly after.

"I don't know …" Dorsey looked around. He probably still thought they should be in Maui.

This felt right. "The Centre is a scientific enterprise; public, this says. So not under corporate scrutiny. I bet there was someone there they knew: either from postgraduate school or met at a conference or something. They could have helped move and hide the safe." Mike and Dorsey finally got to the top. She started back down the path. "It all still matches the message: Puff's home, a cave: the C.A.V.E. Under the sad room—those lights are just like SAD lights, but for plants. It's as good or better a match than here. Come on!" How come they were so slow?

Dorsey and Mike followed her to the car, a downhill walk that seemed immeasurably longer than the uphill one they had just accomplished. Once they were all in, Dorsey turned on the engine and the radio, then said, "Let's maybe be a little less enthusiastic at the aquarium."

"Oh?" said Mike, with a leading tone.

Dorsey cleared his throat. "I was fairly sure no one was following us up to the cave. I can't be so sure at the aquarium."

"But these are your people?" she asked.

Dorsey put the car in reverse. "I trust the people in this car."

She leaned back in her seat and folded her arms, some of her excitement draining into worry. If it was in such a public place, would they be able to get at it unseen?

* * *

Save the C.A.V.E.! the banner on the stairs read. If they had just left a little later, or Carolyn had actually looked at the flyer, perhaps they would have saved themselves a trip. Although would she have noticed, without the disappointment of the reality of the blue room? It didn't matter. They were here now.

They started up the left-hand stairs. She paused at the landing on the cafeteria level, where you could see both the vivariums and aquariums. The fuchsia light bathed the left-hand side. Her eyes caught on the shipwreck scene below it: objects strewn about, just like in the Paris junkyard. "No. Down." She turned around. Her parents had made a more clever message, but not changed the code—perhaps they had also not changed the type of the place they thought was good for hiding.

She walked up to the aquarium wall. A turtle leisurely swam by. A school of bright blue fish darted in and out of an artistically broken ship. An eel popped half out of a metal tube, then retreated. It came out again, an undulating snake, and a school of tiny iridescent fish split around it.

She shook her head. She was meant to be looking at the display, not the fish. The scene recreated an accumulated graveyard of ships: to her left, a broken wood deck showed iconic treasure chests, teasingly part open with shiny gold sparkles from inside; directly in front, a smoother, blue-and-green sailing boat with a tangle of thin broken masts and spars; and to the right, a shiny, silver modern-looking vessel, with bulbous windows. Detritus scattered around each ship. The old-fashioned ship had more treasure chests beside it, plus ancient-looking mariner-type tools. Sextant and telescope she could identify; others were more obscure. The middle boat sat amongst broken deck chairs and several leather-looking briefcases with combination locks. The modern ship was surrounded by various metallic lumps, some shiny briefcases and others larger. She walked right, examining these. Corral had accreted—or been sculpted—on top of most of the objects, obscuring their outlines. But one was sharply boxy, like a cube. It sat in the centre of a patch of tall seagrass. Unlike every other lump, it did not have some kind of artistic damage.

She moved farther right, trying to get a look around what seemed like it would be the 'front' of the piece. The school of blue fish swam past, causing the grass to temporarily swish out of the way. She caught glimpse of a stylised DNA helix before the grass swished back.

She felt filled with energy; blood pumped to her head. She lifted an arm to point, then remembered Dorsey's comment in the car. She changed the gesture into patting her hair. "There," she whispered.

Dorsey and Mike joined her. "Wow," said Mike softly. "How could no one have ever noticed this?"

She gestured across the whole display. "There's 'treasure' everywhere. I'm sure they did. But it fits right in."

"Genius," said Mike. She smiled, oddly proud of her parents for gaining the praise of a spy, decades after their death.

They continued to stare in silence, perhaps all caught up in their internal worlds. Her heart wouldn't stop thumping a fast staccato. She wanted to jump and shout and run. But Dorsey's words echoed in her mind, and she only leaned forward for a better view.

She could not believe she was standing some three metres, one sheet of hardened acrylic, and several tons of water away from what had to be her parents' secret. This safe wouldn't be empty. They had hidden it ahead of their planned arrival, then gone on their holiday—and been murdered. Whatever they did right before they died had been sitting in this aquarium for the last three decades.

How would they get at it? She frowned, the reality of the problem sinking in. Her parents would have had their friends at the aquarium, willing to help. She and the agents were just interlopers, unwilling to do anything too dramatic for fear of attracting attention. But perhaps they could come at night and dive?

"I think we might need that diving gear …" she said, just as Mike said, "Can you really get diving …"

They looked at each other and grinned. She wondered what chain of thought Mike had followed, that took exactly the same amount of time to reach the same point she had.

"I can get diving gear," said Dorsey softly. "I just need to think a bit more about what to say about why I need it."

* * *

She stood, uncomfortable, in the hotel shop, holding a flowered bikini—the only swimming costume in the place that fit her. "It's terribly expensive."

"Don't worry," said Mike with a smile. "The US government is footing the bill."

"I really don't need a swimsuit," she said. "Just you and Dorsey are planning to dive."

"We never know what might happen. Besides, we've got the whole afternoon." He gestured about them. "Here we are, in paradise! How many times in your life are you going to find yourself at a four-star hotel in Hawaii?"

Never, would have been her original thought. "Probably not again."

"See? Don't squander the chance, hiding away in the room. All our dubious friends are gone. We can have a lovely afternoon on the poolside."

"Are you sure?" She meant both about the 'friends' as well as the invitation to visit the pool. "With everything going on, it doesn't seem quite right …" It seemed somehow disrespectful to relax, here, at the final leg of the trail to parents' secret. Her life, Susan's life, and Ethan's life were still in shambles. She had no idea how things could be fixed, even once they found the secret. How could she spend an afternoon swimming?

Mike's expression grew serious. "My whole life is a series of events like this. You take the joy where you can find it." His face relaxed into a broad grin again. "Come on, do it for me—otherwise I'll be cooped up in the room too. You don't *have* to get in the water."

"Okay." She handed the bikini across for him to take to the till.

* * *

She wrapped herself in the soft robe, hiding temporarily from view the overpriced bikini and second-guessing her choice. If the one-pieces hadn't been too small, would she feel more comfortable? She was unsure whether she was feeling strange about squeezing in holiday activities or just exposed in the outfit.

"Ready?" Mike called.

"Yes." She slid her feet into the hotel's rubbery pool slippers.

She and Mike walked around to the pool. Dorsey was already there, swimming seriously along the straightest side of the waving, definitely-not-made-for laps pool. Mike dropped his robe on a beach chair, took two long steps, and cannon-balled into a clear space. She couldn't help admiring his taut, muscular physique. But of course a spy would be in great shape. Strangely, though, she had not had the same thought about Dorsey.

Mike bounced up and stood, waving. "Come on in, the water's great!"

She removed her slippers and set them neatly the foot of the neighbouring chair, then self-consciously took off her robe and draped it over the back. She went around to the stairs and stepped in quickly, relieved to find the water pleasantly warm. Once she was mostly under water, the bikini did not seem so revealing anymore. She leisurely swam a breast-stroke. Mike pushed a floating tube towards her. She linked her arms over the tube, then rolled onto her back and looked up at the sky. She felt more comfortable not looking directly at Mike.

"What do you think it will be?" asked Mike. She turned her head slightly. He was similarly floating on a tube.

"I have no idea," she said honestly. White, puffy clouds blew slowly past in the vibrant blue sky. She thought again about her frightened, stressed parent, hiding some secret only weeks before the end of his or her life. Had they known the extent of their danger? What would her parents have thought of her, here, searching for their secret?

She had never wondered what her parents would think of her before. She had spent so much time thinking about them from her perspective. How their history had burdened her life, first making her a curiosity, then a laughing-stock. But no matter what deception they had generated around her, they were still parents. Uncle Keith had always avowed they loved her. Her few scant memories were pleasant. Her heart clenched, imagining Ellen, two decades into the future, chasing down some mystery Carolyn had left behind. What if Carolyn never returned? What if their last moments together had been that hurried trip to drop her off at Nicholas's?

She thrust the thought violently from her head. Uncle Keith, Bae, and Nicholas would take good care of Ellen. Carolyn would come back, this issue with her parents and Sandslin and whoever else sorted.

*   *   *

Carolyn sat, clutching her backpack, feeling nervous under the gaze of the same anticorporate woman who had handed her a flier earlier in the day. The fishy smell of the aquarium water filled the whole of the small access space, and the lights remained dim. She could barely make out Mike and Dorsey below the water to their left. Dorsey had obtained the diving gear, and Mike the contacts to get them into the C.A.V.E. She wondered who the woman thought they were; she should have asked. She frowned, wondering if Mike's insistence on an afternoon swimming had been a purposeful distraction to keep her from questioning their plans too much.

Yet that made little sense. There wasn't any reason to keep her in the dark, was there? Dorsey's words about trusting just the two of them returned to her mind. As did her own knowledge that she had hidden—was hiding—information from them. Maybe no one really trusted anyone.

Splashing sounded. Dorsey and Mike clambered to the surface, balancing a near-replica of the safe in the Paris underground onto the movable walkway above the aquarium—though covered in algae rather than dust. The crane holding it groaned, and she remembered the slowly crumpling washing machines under the other one. But it was up, and the two agents had dealt with the tricky bit. She just needed to place a finger in the lock.

The walkway slowly ratcheted over towards the side where she and the other woman waited, with Mike and Dorsey stabilising the safe. It finally reached the edge. They used the crane one last time to swing it over onto the solid floor.

She stood and approached nervously. She waited while Mike and Dorsey towelled themselves off. She ducked her head, averting her gaze from Mike. She didn't want him to think she was staring at him.

"Ready to give it a go?" Mike asked.

"Do we, um, need to dry it off or something?" she asked.

"These are built like tanks," said Dorsey. "It shouldn't be bothered by having been underwater for decades."

"Okay," she said, although that was not quite what she had asked. She knelt in front of it. "Has anyone tried the handle?" She didn't want a repeat of last time.

"Actually, no." Mike grinned and gave it a push. "Locked." He tapped at the panels on the front like he had in Paris; one swung back, disgorging a splutter of water.

"Okay." Her stomach fluttered. She took a breath and tried to force her hand still as she reached out. It still shook some. She jammed her left index finger in. She felt the same heat, then brief sharp pain, that she remembered. The lights on the panel began a complex set of flashes: blue, green, red. She removed her hand. Everyone watched in silence. After what seemed like ages, but could not have been more than a few minutes, the lights settled to red and stayed there.

There had not been any unlocking sound. She tried the handle anyway. It moved a few millimetres, then stopped. Her heart sank.

Mike reached in front of her and tried the handle again. "Are you *sure* you're Carolyn Schwarz?"

Chapter Thirty-Six

She stared at the steady red gaze of the lights. They flashed on and off three times, and the panel went dark. "Let me try again."

She used her other hand this time, although her DNA should be the same in both. The lights went red again. She tried the handle anyway. Still locked. "I don't understand."

One of the mysterious anticorporate helpers spoke up. "Carolyn Schwarz was supposed to be genengineered. Maybe you're adopted instead?"

Carolyn pulled out her mother's ID from around her neck. "They found someone who looked just like my mother, if so."

The man leaned forward to squint at the ID. "Wow. Yeah, um, you do look similar."

She *had* to be her parents' biological child. So why wasn't it working? "Are we sure this is tied to first degree? Maybe it's just for the two of them?"

The anticorporate man who had spoken lifted the panel to the left of the one she had used and looked inside. "I think I can get the data here … Yes. It wasn't protected, because opening was tied to the genes. Back then you couldn't synthesize a whole human genome, so it was fine to show what was needed to open it." He snapped his face towards Carolyn. "Um, or at least everybody else thought no one else could synthesize a whole human genome? The safes had been around for a while at that point. There might have been newer ones but maybe they were more expensive …"

"You don't have to defend my parents not acting like they had really synthesized me," said Carolyn.

He tilted his head and stared at her, as if really looking at her for the first time. "Right." He reached into a pocket and pulled out a datastick and a handful of adapter cables. "I think I can get the info out, and we can take a look at it directly. We might even be able to synthesise it onsite." He flipped through his adapters, then chose one. "In the meantime … you're really Carolyn Schwarz? Where have you *been?*"

The last was said in a tone of utter quizzicalness. She realised he would know nothing about her—surely Mike hadn't passed on her entire history. Although she was somewhat surprised he had even passed on her identity. But perhaps he trusted these anticorporate people like he had trusted the French.

The French: she suddenly remembered she had their two mitocyls in her bag; she had unpacked clothes and toiletries, but the mitocyls should still be in the bottom. "Wait, I can *test* if I'm really …" She trailed off. She couldn't quite say it—if she was really herself? Who else would she be? Her stomach felt sick. Her self-image had already been rocked once, as a teenager, discovering she *wasn't* the first fully synthesised human. What if she wasn't even related to her parents at all? What if they had just—somehow—gotten a random child?

She pulled her bag back around to her front and dug inside. She pulled out the mitocyls. "How do these work?" Albert and his friends had just jabbed them at Susan, she remembered. But perhaps there was some kind of activation.

"Woah, is that a mitocyl?" The anticorporate man turned from where he had knelt before the safe and reached out. She handed it across. "For Rebecca Schwarz? How did someone get this?"

Carolyn stared at Mike. Yeah, how had the French gotten that? Where would the spies have gotten her mother's mitochondrial sequence from?

Mike cleared his throat. "She studied in the States. It was on file." So he was the source.

"That's definitely my mother's sequence in there?" she asked. Mike nodded. "Well, let's test me. I *am* Carolyn Schwarz, but we know they lied about where I came from in the first place. I just assumed I was their natural kid …"

Mike reached out and touched her elbow. "Hey, it's okay. You're still you." She smiled, oddly comforted, despite the meaninglessness of his words. "Do you really want to do this?"

"Yes." She wasn't going to come this far and leave it a mystery.

Mike took the mitocyl, and before she knew what had happened, there was a sharp jab in her upper arm, less intense than the one from the safe. He squinted at the mitocyl. "You're your mother's daughter." He grinned.

She smiled back. They locked eyes, sharing joint relief. She wondered briefly why *he* seemed as happy about it as she did. It did not answer the issue at hand, however. "So why isn't the safe opening? Maybe it does need to dry out?"

The anticorporate man stood from in front of the safe. "I've got what I need. Ms Schwarz, um, if you wouldn't mind sparing a few cheek cells, we could compare your DNA with the safe's lock." He folded his hands diffidently, as if he was afraid of scaring or offending her.

He would not know she was a scientist, she realised. Perhaps that was were the hesitance came from. He had just seen her place her finger, twice, into the safe's mechanism and get tested by a mitocyl. It was not as if she was worried about hiding her DNA. "Sure."

She wanted to know what was going on. Maybe the safe wasn't even keyed to her parents. After all, they knew what was inside. Maybe it was for one of their friends, like Ethan, to whom they had been hinting directions for his research. Wouldn't it be ironic if it was Ethan who should have come on this trip? Everyone had had him before her.

The anticorporate woman stood. "Let's get down to the lab, then."

The third person, a man, looked at his two companions and then back at the safe. "I'll stay here and keep on eye on this."

Dorsey mimicked him, looking between Mike and Carolyn and the safe. "Me too."

The two men sat on benches on either side of the open space in which the safe sat, eyeing each other. The anticorporate woman rolled her eyes and gave a small frustrated *huff*. Carolyn felt a sudden sympathetic connection with her. She grinned. The woman grinned back.

*   *   *

Her cheek cells were busy having their DNA extracted in one machine, which would feed directly into the sequencer. They were in a laboratory, packed densely with equipment and computer screens, with another similar, but smaller, room accessed through a doorway to the left of the main computer terminals. She stood with the anticorporate man and woman in front of the largest screen, across which another sequence scrolled by. They weren't looking at the screen, though, and were instead tabbing through an output piped to an elly-book.

"Two full genomes, with the opening sequence specified as to a fifty percent match to the stored sequence," said the man. "So either of them could open it. Or a first-degree relative of both of them."

She tilted her head. "But could my father open it? The Y-chromosome is so much smaller, it wouldn't quite be fifty percent. Or is that within acceptable error?" She frowned, thinking aloud. "Surely any sequencing that happened as fast as that safe responded wouldn't be measuring the whole genome, just a random sample."

The man squinted at her. "The match is on the chromosome level, so it has to be fifty percent of the chromosomes matching." He swiped at the elly-book. "But you're right; the sequence is randomly selected."

"The chromosome level?" She found herself staring at her arm, like Susan had, when the police woman had been trying to imagine all the Barr bodies in each of her cells. Her heart leapt in hope. "That's why it didn't open. My three X-chromosomes would push it over fifty percent."

"But wouldn't it just double up on one of your mother's?" asked Mike.

"No," she said. "Nondisjunction means I'd have both of her two different X's, plus my dad's."

"No, no." The man swiped more. "It's *at least* fifty percent. You should have opened it."

Her excitement drained away. "Oh. So I'm not their kid after all."

Mike brandished the mitocyl. "You are your mother's daughter; we determined that."

She tilted her head. "I share her mitochondria. It could be anyone in the female line. Maybe I'm some other relative's kid. That does happen." Uncle Keith was her mother's only sibling, but there were cousins on their mother's side. They had been much older than her, and she did not really interact with them so much. Who *was* she, anyway? She had a sudden urge to rush to Uncle Keith and demand he answer her. If her parents had adopted a family baby for some reason, he should have known.

The woman looked over the machine showing the progress of processing Carolyn's actual DNA. "We'll know in a matter of hours. What about we all get some sleep, and see in the morning."

Sleep! But the anticorporate woman was right. Carolyn was fuzzy-headed enough now. Staying awake for the hours it would take to get her full genome sequence would make her no better, and she had not spent the first part of the evening doing a complicated dive.

Mike stared at the woman, then nodded, as if making the same calculation. "You have somewhere we can crash?"

*   *   *

Carolyn woke with sunlight streaming through the window of the coffee room, where she, Mike, and Dorsey lay sprawled across the two couches and a chair—Mike, as shortest, had taken the chair. He was still asleep, slumped with one arm and leg over the chair's arms, one arm over its back, and a leg stretching out into open space in front of him. His foot nearly touched the safe, which had been transported down to their sleeping location. His hair lay stretched above his head, looking somewhat comical. She smiled. He did not look like a dangerous spy.

She sat up, and Mike jerked awake. But he did have those spy-reflexes. Dorsey stretched and rolled over. She had slept through her chance to tunnel Susan in the night, but with those reflexes she probably would not have been able to keep it hidden.

Mike grinned at her. "Shall we get to the lab?"

Her stomach rumbled; she was always one for eating as soon as she woke. But finding out what her sequence had revealed was a far bigger draw. "Yes." She stood, then looked back at Dorsey and the safe. "Can we leave him and that here? What about when everyone shows up? It's a working day, right?" She was starting to lose track of time.

"We're good here, I think," said Mike.

"What do they know about us?" she asked as they walked down the corridor.

"It's complicated."

"Well, they know who I am. Or at least who my parents are. Who do they think you are?"

"A spy." He grinned, then headed down the staircase. They were at the lab moments later, and Carolyn was unable to follow up her questioning.

The anticorporate man who had accompanied them to the lab before was there, sitting a computer terminal. The screen showed a familiar sight, a rainbow bar across the top with numbers inside, and then two paired lines: one black, and one incredibly spotty one that was mostly blue and black. He was scrolling slowly down the output showing the paired lines.

"You ran an alignment already?" Carolyn pulled a chair up to sit beside him. "What am I looking at here?"

"I just started. This is you against your parents, chromosome one. I filtered it against standard human-to-human similarity, otherwise it was a complete match."

"Doesn't look good, does it? Is it *both* copies of chromosome one? How far have you gotten?" She would only have one of each of her parents' chromosomes; maybe this was the one she didn't have.

"I'm into your father's second copy. The first was the same."

Her heart dropped. "Oh." She had hoped, against all the evidence of last night, that it would still show her as her parents' child.

The screen turned black and red, with a smattering of pink. He stopped scrolling, then started again, faster. The colours remained the same.

Her heart leapt up again. "That's a match!"

"Your mother." He continued to scroll. "Second copy ..." The screen remained black, red, and pink.

"Am I a *clone*?" That could explain how closely she resembled her mother. Clones had never quite taken off, not like the designer babies. In the end, very few parents wanted an identical twin of one of them. But perhaps her parents had, or settled for it, when their synthetic efforts failed. "But where did the extra X come from?"

"Let's find out. Sex chromosomes are at the end." He pulled the screen all the way to the bottom of the file, then scrolled up. "Mother to mother here … father's Y …" It jumped, briefly, to the spotty black/blue, then back to red/pink. That was a match again. "Father's X."

"The third X is from my father. That doesn't even make sense. How could they have gotten just one chromosome if they cloned my mother? And by *mistake?*" She stared at the anticorporate scientist-man. He stared back, eyes equally puzzled. She shifted sideways, to get more in front of the screen, and he let her. She zoomed in on her father's X-chromosome. "What are all these pink spots, anyway? Why does some of it have lower alignment? That doesn't make sense, either."

She tapped along the side, where the view controls were, turning on annotation. She wanted to know what parts of the chromosome were a perfect match—red—and what parts were a worse match, the pink. The annotation, which specified areas of the genome—either genes, or promoter regions which controlled genes—was so dense as to be unreadable at this coarse view. She continued to zoom in. The red and pink pattern remained nearly constant, fractal-like, until the point when the annotations became readable. Then the pink stayed in place, stretching along the length of each annotated gene.

"The protein-coding regions are the lower match," she stated, unable to do more than narrate what she saw. She kept scrolling, and it kept being the entire length of genes themselves that appeared pink; every once and a while a promoter region was pink, too. It didn't make sense. "How could only some of the sequence match like this? I guess that explains why the safe didn't open." If she had a less than perfect match to her parents' chromosomes, and it worked on a chromosome-by-chromosome basis, she would not hit a match on any of them. "But how could this have happened? Cross-over couldn't mix up the sequence like this. Not so frequent. Not so perfectly on just the coding sequences." She could not think of any biological process which could have left this sort of mangling of her parents' genes in their offspring.

No *biological* process.

No natural way of generating the sequence—her sequence.

Wonder filled her. She dropped her hands and pushed away from the terminal. Her chest felt as if it would burst; her head seemed to be floating. "Oh my God," she said. "I *am* synthesised."

Chapter Thirty-Seven

Everyone stared at the screen in silence. Possibly with their minds full of the same thoughts as Carolyn's: *No, it couldn't be? Could it? How* else *to explain this pattern?*

They had only looked at two chromosome pairs, number one and the sex chromosomes. What about the rest? She scooted back to the screen and started scrolling, but it was too slow. "Is there a …" Her questing taps pulled up the navigation menu. She zoomed out, then jumped from chromosome to chromosome. The twenty-one other pairs were same: no match to her father, near-exact match to her mother. She was an almost-clone of her mother.

"Why did they give me my Dad's X?" It didn't make sense.

"To make it look like you weren't synthesised?" suggested Mike, who seemed a bit less stunned than the scientists. "Maybe they had planned the reveal sometime soon, to protect you."

"But why would I need protecting? I was a novelty, but not much else. I made it to sixteen thinking I was synthesised before I found out the truth …" She trailed off. It wasn't the truth after all, was it? Her brain had trouble wrapping itself around this new reality.

"Maybe they thought you could be a target, based on whatever these Final Findings are?"

"But *how*? I was already supposedly … actually … a scientific breakthrough. Just because the people who created me also found something else wouldn't make *me* a target."

"We still don't know what they found," said Mike.

"Good point." Carolyn pressed her fingertips to her forehead. "And now we won't. I can't open the safe."

"We can synthesise the opening sequence," said the woman scientist. "I'll get started on that."

"How long will that take?" asked Mike.

"Two or three days."

Mike frowned.

"The human genome is large," she said, eyebrows raised.

Carolyn asked, "Do we need to synthesise the whole thing? If it only takes a random sample, we could just present less."

"That's a good idea," said the man. "I'll check what the sampling regime is." He picked up the same elly-book from the night before, made a low *hmm*, and retreated to the small side room. Mike and Dorsey followed him.

Carolyn leaned back in her chair, too emotionally exhausted for the moment to join them. She closed her eyes, letting senses wash over her: the sounds of the men in the next room, the rough fabric of the chair against her back, the faint fishy smell that pervaded the whole C.A.V.E. She opened her eyes and focussed on the woman. "So, um, where's your other friend from last night? And what time is it? What about when everyone gets into work?"

The woman plopped down into a chair. "We're all that's left."

"To run this whole place?" Carolyn asked.

"We've got volunteers who do maintenance—mostly kids and college students. The tourist stuff is still run by the state. They handle entry and the cafe. We do everything else."

"Wow." The C.A.V.E. did need saving. "So they want to sell it to a corporation? Why?"

"I don't know." The woman turned to the look at the screen, which still showed Carolyn's sequence compared to her parents'. "I'm not sure they do. The real question, is why do the corporations want us?"

"Why ..."

"I don't know that, either."

An uncomfortable silence lengthened. Carolyn said, "Hey, we haven't really been introduced. I'm Carolyn." She smiled. "But you knew that already."

"I'm Leia," said the woman, straightening, as if she were relieved Carolyn had started the conversation again. "My friend there,"—she gestured towards the other room—"is Bane, and George is the one who keeps eyeing the safe. Bane and I are fish ecologists; George is an aquatic botanist."

"So, um, how do you know ... why are we ..." *Why are you helping us?* she wanted to ask, but did not quite know how to phrase it.

She shrugged. "Mike asked us." She shook her head. "I can't believe the Schwarz Final Findings have been sitting in a display here this whole time." She leaned back. "I mean, lab lore says the Schwarzes knew our boss's boss. There are photos of them in D.C. as graduate students. But that's well before they became famous. That they could have hid the Final Findings *here* ..."

"We don't know that's what's there." But what else could it be? Leia hadn't really answered her question. "So, you know Mike from ...?"

"That isn't that necessary, is it?" Mike interrupted, leaning around the doorframe.

Carolyn spun to face him. What was he trying to hide? Anger flashed. She was tired of being left in the dark. By Mike. By her parents. By everyone. "Perhaps if you just spoke straight once and a while it wouldn't be!" She found herself standing, towering over him. She sunk back down into the chair. "Sorry. I just want my life back." But she didn't, anymore, did she? She couldn't go back to her old life, doing research supported by corporations who were far more nefarious than she had ever imagined. "Is there any independent research left?" she asked, voicing her thoughts.

"We are," said Leia. "But not for much longer."

Carolyn stared at her. "That's why they want you."

"What?"

"That's why the corporations want you. You're actual, independent research, without a corporate sponsor?"

"Yes." Leia's voice rose hesitantly.

"Do you realise how rare that is? The corporations own all the research, in Europe at least. I can't imagine it's that much different in the States. Most of the corporations *came* from there—here—in the first place."

"I've told you that you work in a bubble," said Mike.

"A collapsing bubble." Leia tilted her head and stared at Carolyn. "How do you know so much about research?" She squinted at the screen. "How did you know how to use the genome browser?"

"I'm a scientist." Carolyn gave a crooked grin. "I didn't mean to be …"

"How come we've never heard of you?"

"I use a different name. Schwarz was too …" She couldn't think of what to say. Maybe if she hadn't tried to erase her past as a teenager, would being the child of the Schwarzes be so bad? Yet would she ever even have made it into science with that hanging over her? All these were questions about a past that didn't happen.

"Yeah, I can see that," said Leia. "So, what do you study?"

"Mitochondria," she said.

"Like Ethan Boltzer? He was one of their crowd too."

"Well, not quite like Ethan Boltzer. We use the Boltzer process, though." She looked at Mike and did not add that she had met Ethan. She did not know who knew what. She wished Mike wasn't being so secretive.

Carolyn bubbled with amusement; a spy was secretive nearly by definition. She stopped short the laugh that was fighting its way out. She was on her way to inappropriate hysteria. They needed another topic of conversation. She looked at the screen again. "I wonder what the differences between me and my mother *are*. Why did they change the coding sequences? That's so strange."

"They could not have changed them too much," said Leia. "You're her spitting image."

Carolyn scooted up to the screen. "How do you get to the raw alignment?"

"It's over here." Leia pulled up the navigation menu and tapped options along the bottom.

Carolyn saw what she needed to do. She pulled the screen to a coding region on her mother's fourth chromosome, which was the expected pink on her matching sequence, then tapped the alignment icon.

The screen changed from colours to plain letters. Paired rows of letters—the A, C, T, and G representing nucleotides of the genetic sequence—filled the screen, connected by short vertical lines where the sequence was the same between her and her mother. A clear pattern jumped out immediately: the sequence matched for two out of every three nucleotides, nearly consistently. "What?" She scrolled up to the start of the gene. The first five nucleotides matched, then the pattern began. "Well, there's only one codon for methionine, so that had to match," she said out loud. Otherwise it wouldn't be the start of a coding sequence, the cue that she and Ethan had noticed on the ID cards. "But all the rest of the amino acids …" Each codon was one nucleotide different from her mother's. She scrolled along. Almost—every once and a while a set of five matched in a row, or the pattern jumped about briefly. "What's the amino acid match?" She found the appropriate controls and swapped to a view where the letters were now the same alphabet as had made up the code on her parents' ID cards.

Now the paired rows of letters were completely connected by vertical lines. She had the same amino acid sequence as her mother, just not the same DNA sequence. "They used alternative codons? Whatever *for?*"

"Maybe it was meant to prove you were really synthesised," said Leia. "Otherwise how could they distinguish you from a clone of your mother? Clones were old tech by then."

"Then why did they add the extra X, making me look like I *wasn't* synthesised?"

"Maybe they were covering all bases?" Mike came fully into the room. "They could have gone either way, I suppose. Irrefutable evidence that you are synthesised—is that truly the case? You're absolutely, definitely synthesised?"

Carolyn switched the screen back to the DNA view. "This could not have happened by accident."

Leia nodded. "Not even some massively error-prone PCR of your mother's sequence could possibly make so many errors that just happen to be codons for the right amino acids and just in the coding sequence."

"Okay," said Mike. "So either irrefutable evidence that you were synthesised, or the circumstantial evidence—that everyone believed—that you weren't. Neat." His voice held that same tone of appreciation it had earlier, when they had first seen the safe. He cleared his throat. "So, um, can you explain what's going on here, then? You're a clone, but not quite?"

"That's just about it," said Carolyn. "You said you knew about the codon code?"

Mike nodded enthusiastically. "It's degenerate. Three letters—or DNA letters ..."

"Nucleotides," supplied Carolyn.

"Three nucleotides code for one amino acid ..." He paused, as if waiting for confirmation of his terminology.

"That's right."

"But there are more amino acids than sets of three nucleotides, so multiple codons code for the same one. You told me about some kind of other DNA? tDNA?"

"tRNA," said Carolyn. "That's what reads the code. Each codon has a tRNA that goes with it—RNA is like DNA, it has nucleotides, but they have a slightly different chemical structure. So it can match up to complementary DNA, just like DNA can."

"Um, I don't think you really explained this bit," said Mike.

Carolyn straightened, teacher mode coming well into the fore. "Do you see all these letters?" She gestured at the screen. "Each one represents a nucleotide, and the DNA sequence is made up of the string of them put together. Here's my mother; here's me. But DNA isn't just one string of nucleotides—it's two. But we only write out one, because that's the key feature of DNA: each nucleotide binds only to one other nucleotide. We call each string of nucleotides a strand, and the two strands are complementary. Whenever there is an A on one strand, it will bind to a T on the other, and whenever there is a C on one strand it will bind to a G on the other." She ran her hand over the screen, pointing out the letters in question. "So we just need to write one strand, since that specifies the other. When DNA copies itself, it separates the two strands and builds another complementary strand for each. That's how DNA sequence is passed down from parent to child."

"Okay, so what does that have to do with this codon code?"

"Right, anyway, so the RNA does this too. When a gene is being read as a protein, the first thing that happens is that the DNA opens up a bit, and what's known as a messenger RNA is made: this is an RNA copy of the DNA coding sequence, using the complementary letters. We call it mRNA for short. Then this mRNA is read by a set of cellular machinery known as ribosomes, which bring the mRNA in contact with the tRNAs—transfer RNAs. The tRNAs have a bit that's complementary to a codon on one end and a bit that binds to a particular amino acid on another. There are tRNAs for each codon except the three stop codons. So a protein is built one amino acid at a time, as a tRNA matches to the codon and drops off its amino acid, then the ribosome moves onto the next codon."

"What's going on here?" He gestured at the screen.

"For some reason …" She trailed off. Was it just as simple as hedging their bets, planning to either prove she was synthesised or suggest she was not? Or was there another point to making *nearly every codon* different from her mother's? That wasn't explaining to Mike. She shifted back into teacher mode. "Instead of just synthesising my mother's sequence, they replaced almost all the codons in my mother's genes with an alternate codon for the same amino acid. So I'm a clone on the protein level. Functionally, my mother and I would be the same. The DNA is just a code for the proteins, and as the amino acids match, the proteins are the same."

"Just made by a different code," said Mike, in a tone of wonder. "You and your mother are like the ultimate example of a degenerate code. Two different codes, that when decoded, produce the same message."

"I suppose so." Carolyn was not used to thinking of herself as either a 'code' or a 'message'. But clearly that was how a cryptographer spy thought.

Bane and Dorsey emerged from the side room. "I've got it!" Bane said. "It randomly selects sequences and reads sixty base pairs at a time. It does this for only about a hundred thousand reads. So we only need to make one to two hundred thousand short sequences—I'd suggest one hundred fifty base pairs, to make sure it can get up to sixty if it grabs it in the middle—and scatter them across all the chromosomes."

Leia stood up. "On it! That shouldn't take long at all." She grinned. "In about twenty minutes we can go open the safe."

Chapter Thirty-Eight

Carolyn couldn't remember what she had eaten for breakfast, just that it had assuaged what had been a sudden ravenous hunger. And had taken just about as long to make and eat as the DNA sequence did to synthesise.

George appeared from somewhere, and the six surrounded the safe again, in the coffee room. This time it was Bane who approached it with trepidation, sticking a wobbly, gelatinous rod into the flap where Carolyn had presented her finger.

Nothing happened.

"It's not taking," said Bane.

"Try the felt." Leia tapped at the elly-book. "Warm it up first."

"We'll break up the DNA," said Bane.

"Warm it up *before* you soak it in the DNA," said Leia.

Carolyn clenched and unclenched her fists. Maybe she could just dip her finger in the solution …

But Bane was already removing a small piece of felt from the microwave. He flipped it back and forth between his hands, then dipped it into a small tube of liquid.

He used a stylus to balance it into the hole in the safe. Success: the panel began flashing the pattern of red, blue, and green lights. Carolyn held her breath. *Don't end on red,* she said to herself. *Don't end on red.*

The lights turned green, and there was a soft clunk.

It had worked.

They stared at the safe, then each other. Carolyn reached forward and pressed down the handle. It caught after a few millimetres. A tiny spark of fear started in her chest. But more pressure, and the handle moved the full motion. Relief flooded her. She pulled the door open.

The cavernous centre of the safe looked empty for a terrifying moment, then Carolyn's eyes resolved an elly-book—an *old* elly-book—on the floor of the safe. It was the same grey as the interior, and larger in dimensions, yet thinner, than modern elly-books. Before docfilm, Carolyn thought to herself, when elly-books were also trying to capture the 'thin' market taken over by the paper-replacement.

Everyone seemed to be waiting for Carolyn to take the lead. She reached in and lifted the elly-book. It was unexpectedly heavy. She pressed buttons, but nothing happened. "It won't still be charged," she said, slowly realising the obvious. Even a modern elly-book would not retain a charge for decades, much less this ancient model.

Mike stared at it. "How can we charge it?"

"Never fear!" said Bane. "We've got a century's worth of cables in the drawers here. We'll find something that fits."

* * *

The elly-book sat attached by a daisy-chain of three cables to a power socket, beside the screen where they had viewed the alignment results. Carolyn stared at it.

"A watched device never charges," said Mike.

"I don't think that's the saying." But Carolyn smiled. His words were having what was probably his desired effect, to distract her.

"I really don't think boiling it would be a good idea," he replied.

She laughed out loud. It wasn't even that funny. But she was so keyed up, everything seemed just ... more. "What will it say?" What could possibly be so incredibly amazing that people were searching after it decades later? Her parents *had* synthesised her. They weren't washed-up frauds; they were scientists decades ahead of their time. What else might they have found?

She shook her head. She needed something to distract her. She turned to Leia. "Can I get a copy of those genomes?" Even though it was just a long sequence of nucleotides, her parents' genetic code seemed yet another connection to the real people. And whatever they had done to hers ... That deliberate changing of the codons was an ingenious way to prove her synthetic origin. Not that it had ever gotten that far. Covering their bases, Mike had said. She could be revealed as a hoax—as she had been—or real—as was hidden. What a strange, frightening world it must have been for her parents.

"Sure." Leia frowned and looked around. "Do you have an elly-book? We probably shouldn't, um, pipe it to the cloud."

"I do!" Carolyn pulled the student's elly-book from her knapsack.

Leia took it. "We'll need to take it to the ports." She walked into the side room Bane had done his work in earlier, and Carolyn followed.

The thought of having her very own copy of her parents' genomes filled her with a strange excitement. But why were Leia and everyone being so accommodating, so easy with them? Except for George, she supposed; he still watched them suspiciously. "How *do* you know Mike," she asked quietly, aware of the others just around the corner.

Leia found a cable and plugged Carolyn's elly-book into a port in the wall. She tilted her head, acting as if she was considering something. "I'm not sure ..."

"I'm tired of secrets," said Carolyn. "Surely it can't be more of a secret than the Schwarz Final Findings?"

"His sister," Leia said. "She's one of us." A scientist? Anticorporate? Carolyn wasn't quite sure what that meant. "I think Mike doesn't like emphasising that, but I've known him since Jenn and I were kids, and he was her too-serious older brother. Jenn even said he didn't like the growing control of the corporations, but he moved into the government. I guess maybe he thought that was better. But I'm not sure."

'Us' meant anticorporate, Carolyn decided. "So is Mike anticorporate?"

"Shh!" Leia checked the elly-book. "Not openly. He can't be. But he keeps in contact through Jenn, and I think it's more than just so he can keep his little spy-network going." She unhooked the elly-book. "All loaded. I've given you the apps, too."

Carolyn followed Leia back into the other room. She wanted to pump Leia for more information, about Mike and his sister and Mike as a child. But this wasn't the time. Maybe there would never be the time. She found the genescan app and opened it, pulling up the genome alignment again. She zoomed in, frowning at the strange pattern of exact matches except for those (usually) final, wobble bases. The pattern that showed, unequivocally, that she was synthesised. But it could not have been as simple as just swapping out codons to design her.

"What are you looking at?" Mike sat beside her.

She tapped the screen. "This. What I said before, that my mother and I should be functionally identical, isn't quite true."

"Oh?" He sounding honestly interested.

She shifted. She wanted to ponder this on her own, and Mike's questions felt annoying. But teacher-mode pulled at her, and perhaps she could work through her thoughts out loud just as well. "Do you remember what I said about codon optimisation?"

He nodded. "Some codons are more common than others. This differs between species. So if you're moving a gene from humans to yeast, you need to switch codons otherwise your gene gets made too slowly."

The fact that he had actually remembered what she had said—how many days ago had it been? Time was blurring together—warmed her heart. She smiled. "Yes. So, if my parents changed my codons ..."

"Some of your genes should be made slower than they should be."

"Yes. Or faster, if it was meant to be slow. I wonder if they modified the frequency of the tRNAs, as well, to make the new code match up." She tapped her fingertips together. "That would be the simplest thing to do. Just swapping the frequency of each tRNA with another for the same amino acid would enable them to match like with like: they'd just use the new most-common codon where the natural most-common codon used to be and so on." She straightened, pleased at having figured it out. She plugged in a search for tRNA sequences, aligning hers and her mothers'. They were different. She picked one amino acid and counted. No, there wasn't nearly so simple a swap. Her excitement faded. "That's not it." She chewed her lip. "But they did do something. My tRNA frequencies aren't natural, either. They're just ... different."

"So it would help fix the speed of the genes?"

"Yes, but not exactly enough to make sure I don't have some massive genetic problem." She remembered how some promoter regions were different as well. She started scrolling to one. "But if they changed ..."

"It's charged!" That was Dorsey, who had obviously spent the time watching the elly-book. "It's turning on!"

She and Mike rushed over. She held her breath. This was it.

The screen flashed blue, then white, then resolved to a pleasant oceanic background with just two icons—files—resting on it. Carolyn again took the lead, tapping the first icon. It opened, and text filled the screen:

To Whomever May Be Reading This: (Ethan? Patty? those are the only names they ever mentioned)

It's happened. I can't believe they're gone. I've done what they said. Although they didn't say to leave this note—but I felt I couldn't just do it and leave nothing. If you've come this far, found the safe, and opened it, surely you deserve some kind of explanation.

Like they asked, I erased the elly-book and overwrote with the C-genome. Don't try to find traces of what was there: it's been overwritten with zeros first, a few hundred times, and then the C-genome written and erased, offset, a few hundred more. There won't be any trace of whatever it was they had here.

Don't try to come find me, either. I didn't look, I swear! I feel I need to add that, because of what has happened. I couldn't look anyway. It was protected by another of their weirdo codes.

So, I'm truly sorry. You must have followed one of the messages that I was not able to find and destroy—they left just too many for me to do that. But with their death, it's clear they were right: whatever they discovered is too dangerous to be out in our lifetime. Science always progresses. I'm confident it will be found again, in time, perhaps in a world more suited than the one we live in.

They were the best bosses I ever had. I'm scared for the future, for all of us, if this is what can happen to honest scientists.

Signed,

A loving lab member. I miss you, Becca and Manny. So much.

Nooooo! Carolyn's mind screamed in despair. It was gone. Whatever it was, whatever had destroyed her life, and Susan's, and Ethan's, for the want of it— didn't exist after all. She moved, zombie-like, closing the file and opening the other.

It was a sequence. The C-genome. C? Baby C. That would be her. Which they already had, with them, in herself, and conveniently sequenced into another, similar file just last night.

"Go back to the letter?" said Mike, subdued.

*   *   *

The rest of the day passed in a haze. Carolyn thought she ate lunch. She vaguely remembered dinner. Mike had said things, about next steps. They knew who Ethan was, but who was Patty? Then there was the letter writer. And anyone else in their lab, if they could be traced. Maybe *somebody* knew *something.*

But it seemed pointless. She had thought there would be an answer—whatever all these corporations were after. But maybe there was nothing. She'd live the rest of her life a shadow, unable to be herself. Unable to be with Ellen? Her heart clenched. Maybe she and Ellen could make a new life together, in one of Bae's underground communities. But that was not fair on her daughter.

She lay awake in the hotel bed, blinking back tears. She clutched the cold metal of her elly-book—the students'—thinking perhaps she could tunnel Susan. But she didn't want to share the despair just yet. She should, before they showed up in person. Although that would reveal her secret communication to Mike and Dorsey, when Ethan and Susan already knew they'd found nothing …

It was just too hard to even think around all the implications. She didn't care, at the moment, if the agents realised she had not shared everything. She had been so *sure,* so sure they would find something. So sure there would be a solution that would fix her life.

At least there had been that note. If it hadn't been there, what would they have thought, finding an ancient elly-book, holding only her genome?

She sat bold upright in bed. *What* would *they have thought?*

Ethan's distressed words, when they had first met, replayed in her mind: *"They told me to look to you for the answer. That they'd passed it on to you."* They'd *passed it on.* Passed it on—as in genes?

Pieces fell together in her mind: her codons had been replaced with alternates, but not by a simple changing of their frequency. She had not gotten to explore, but she knew some of the promoter regions had been changed, too. Perhaps to fine-tune the expression-level of her proteins, when the alternate codons used modified their expression too much. Why *hadn't* they just used the simple method? It would have proved her synthetic origin just as well. There must have been some reason. Some reason they needed to use precisely the codons they did use.

Her parents liked using codes, according to the letter writer. Mike may have been more right than he knew, with his earlier words. She *was* a code.

The idea settled into her. It felt right. It felt more right than anything yet had. She leapt from the bed with the force of her excitement and rushed into the other room. "Mike! Mike!"

Mike sat up, as did Dorsey, both training guns on her. She froze. Perhaps she should have considered a bit more the consequences of startling a spy and a federal agent in their sleep. They blinked, and both slowly lowered their arms. "Carolyn?" asked Mike, concern in his voice. "Are you okay?"

She bounced on her toes. "I'm fine! I know it! I have the answer!" She was being incoherent. "Mike, it's another code. The codons. That's why they've changed them. *I'm* the code."

"What?" asked Dorsey, sleepily.

"Think about it—that note, the writer even said they had not been asked to leave it. So what my parents intended was the safe to hold my genome. What would we have done—what would anyone have done—if they followed all these codes to find a giant DNA sequence on an elly-book at the end of the trail?"

"Look for another code," said Mike, in a wondering tone.

"So, codebreaker, let's look!" She grinned wildly. He grinned back.

Chapter Thirty-Nine

Mike set up his computer in her room with its code-breaking software and loaded on her genome from her parents elly-book; Dorsey went back to sleep. Carolyn settled next to Mike in a chair at the round table near the window. Night sounds, chirping and soft bleeps—frogs?—filtered in. The distant rumble of the waves on the beach seemed louder than in the day.

"Where do we start?" Mike said.

"You're asking me? You're the code-breaker." Carolyn still felt giddy with excitement, the contrast with her depths of despair minutes ago making it seem even more wild.

"You're the geneticist. Where in your genome would you find a code?"

Despite the excitement, her brain was still exhausted from the day and its lack of sleep tonight. So tired, it took her several moments to mentally protest, *Not a geneticist!* She rubbed her fingers together, trying to ground herself in the present. "You said it yesterday. It's in the codons. It has to be. They did something more than just prove I was synthesized. Plus, that's the only thing that's changed, other than a few promoter regions. And the tRNAs." She added the last to be exact. That could be important for a code.

"We'll start with the assumption it's just the codons—the others could be explained by needing to make you work right, correct?"

"Yes."

"So, um." He squinted at the long series of A's, T's, G's and C's on his screen. "Where are the codons?"

She straightened. "In the coding regions. You need an annotation on that." She scooted her elly-book around to in front of her. She opened the genescan app that had her DNA sequenced from the C.A.V.E. "Actually, we should compare these, just to make sure ..." She trailed off. That she was really her? She felt ridiculous.

"That they didn't have a ton of genomes?" he asked. "Maybe you're A, not C. Or maybe you're Z. Good point."

She relaxed, confidence bolstered that he took her seriously. She handed him the elly-book. He pressed it back, handing her instead a long cable.

"Let's hook up together. I'm going to need your input throughout this, so we need to get connected." He plugged one end of the cable into his computer, and she put the other into her elly-book.

She tapped at the screen, readying it to compare her genome with another. "Can you pipe that over to me? I'll take a look in genescan." The genome arrived, and she shifted it to display. "Oh." About half the sequence was the black of a null-match.

"Does that ..."

"No!" she cut him off, realisation dawning. "It's just the order." Half the sequences were black, but neatly divided along chromosomes. When they'd sequenced her genome, it had randomly ordered her paired chromosomes, and the sequence from her parents' elly-book had the other chromosome first, about half the time. "Each chromosome is in a pair, and while we put them in order of one to twenty-two and then the sex chromosomes, there's no reason the chromosome we sequenced first would be the same one they synthesised first." She selected and flipped the order of the ten chromosomes showing black matches and was soon staring at screen full of the red of perfect matches. "Now just let me pull out the coding regions for you." A few more taps, and she had extracted all the human genes into a separate file. "Got them. Ready?"

He nodded distractedly, typing away. "Send it on over!"

She grinned at his bent head. This was exciting. It reminded her of sitting with PhD students or postdocs, chasing down curious findings or getting the results of computer simulations. They were going to crack her code.

*　*　*

Dorsey's voice interrupted a dream where she was swimming after a long string of DNA that was curled around and playing a ukulele. "Found anything yet?"

She sat abruptly and rubbed her forehead, which had creased where it lay against the edge of her elly-book. Mike sat as she last remembered him, hunched over his laptop, typing furiously.

"Getting there," said Mike, with an air of distraction. He templed his fingers and leaned his chin on them, eyes flicking across a scrolling screen. "So close," he murmured. "Carolyn …"

"Yes?" she asked

"I need some more ways to order the degenerate codons. How would you do it?"

"Alphabetically?"

"That's what I'm using now. It doesn't seem to be working."

"What about alphabetically by their complementary sequence?"

"What's that?"

"The sequence that the tRNA would be carrying—so for AUG for methionine for example, the tRNA actually has UAC, because U and A and complementary, and G and C are."

"Wouldn't that just be the opposite order?"

"Not necessarily …" She frowned. "I think." She could not quite get her head around it.

He typed. "No, it's not. I'll use that. What else?"

"Um, GC-content? That's a typical value in biology, although not so meaningful for just three bases."

"I'll give it a go. But something that would have less ties …"

What else did they know about the codons? "Wait! Exactly what we noticed already—their frequency. You could do it by their frequency: in my genome, or what their frequency ought to be. Or the same for the corresponding tRNAs."

"That sounds good." He started typing again. "Good enough I think I'll start there—it would match what we've seen from their codes. They seem to be doing things in themes, and this is already themed on the codon frequency, which they had to modify." He typed more. "I can get the codon frequency in the genome from what you sent, but I need the others …"

"I'm on it!" She activated her elly-book.

"I guess I'll go get breakfast," said Dorsey. She had forgotten about him. He stared at the two of them with a slightly pained, slightly amused, long-suffering look. She wondered if watching Mike absorbed in his code-breaking was something he was used to. It felt nice, somehow, to be included in the same gaze.

"Sure." Mike waved his hand. She turned back to her elly-book, grinning. She was actually having fun, for the first time in this whole adventure.

Chapter Forty

Dorsey burst back into the room. "We need to go, now!"

"What?" Carolyn said. Mike was already moving, slamming his laptop shut and sweeping things off the table into his bag. She stood and followed suit, shoving her elly-book with the now loose cable into her knapsack.

"Sandslin is on their way back here. Maxtech and your friends are sure to follow. We need to be *gone*." Dorsey was already in the room next door, and in no time was waiting at the door with a neatly packed bag. How did he *do* that? She was still standing with her knapsack containing nothing but two mitocyls and an ancient and a modern elly-book.

Mike disappeared into Dorsey's and his room, and she made herself move. Clothes—at least she had changed from her pyjamas sometime in the code-breaking night. She swept yesterday's clothes from where she had left them on a chair into the knapsack and arranged them to cushion the electronics. She shoved other clothes from the dresser in, then her tooth and hair brushes from the bathroom, along with her toiletries. It barely fit—but the only new thing was her parents' ancient elly-book. Susan was clearly a superior packer to her. She shook her head at the random thought.

"Ready?" Mike was back in the room. He took her arm and lead her through the door.

"Yes," she answered, belatedly. "How ... What ...?" *Is your life always like this?* she wanted to ask. She remembered having a similar thought fleeing with Susan. She couldn't remember when. She shook her head again; remembering that was clearly irrelevant to the current flight. With so much to worry about, it seemed what used to be a tendency to worry was now just a tendency to wonder about completely inconsequential things. Perhaps that could be a lesson to her, back in a normal life, when she started worrying. Yet another irrelevant thought. She grinned at herself, her internal amusement somehow relaxing her.

Dorsey was ahead of them in the half-corridor, taking into his wrist. He looked back. "Take an autocab to the airport. I'll meet you there."

* * *

Carolyn started to ask a question once in the autocab, but Mike shushed her, looking at the screen of one of his devices.

"Kalapaki Beach," he said out loud, then typed furiously on his device.

She sat stiffly. There must be surveillance. She hugged her knapsack on her lap. They had been almost there. She was desperate to know what secret her genome held. She opened her mouth to make a comment related to the code, then snapped it shut. That would not be appropriate either.

The autocab delivered them to the airport terminal, Arrivals, despite Mike having done nothing further she could see to their destination. They stepped out of the cab.

He looked at the building. "Close enough."

They wound past the baggage claim, several gardens, and through a flower shop. Mike stopped to buy two breakfast wraps; he handed one to her. She nearly inhaled it, starving. Finally, they emerged near the central check-in desks.

"What now?" she whispered.

He sat on a bench beside an indoor garden. "Wait for Dorsey. Look regular." His voice was pitched low, but not ridiculously so.

"Is it okay to talk?"

"Noncommittal chatting." He leaned back and closed his eyes. "Or get some sleep."

"Sleep?" But just like Susan, Mike seemed to be able to turn off on cue. She leaned back, but didn't bother trying to close her eyes. Her mind spun. Sandslin was on the way back; they must have figured out they had read the message wrong. Would they get it right this time?

What about the other corporations? And Mike's ... colleagues? They had figured out the relation to the Puff dragon song and the North Shore connection. They would probably head to the blue room cave like the three of them had. But would they then think of C.A.V.E.? Would Leia and Bane and George be okay? Had they put the safe back? Would someone remember it was out of the display for a day?

She was worrying too much; she was worrying about other people to avoid worrying about herself. She had no idea how much danger they were in.

Dorsey finally appeared. Mike's eyes snapped open before Dorsey even said anything. Dorsey tilted his head, and they followed him, silently. She couldn't tell if they were being silent because they *had* to, or because they just didn't want to talk. And she couldn't talk to ask, in case it was the first.

She hated this spy stuff. She wanted to get back to the code breaking, which was sort of like science. She wanted to see what her parents had found, which was definitely science. She missed her regular science; she wanted her life back. The familiar refrain fell flat. She didn't want it back anymore, did she?

She set her jaw. She wanted science to be public again, for real. She wanted corporations out of control of the world. She hugged her knapsack tight as they approached a security gate with no other passengers in sight. She was an anticorporate radical, wasn't she?

*   *   *

Neither Dorsey nor Mike spoke until they climbed a set of rolling stairs into a small blue jet. The inside did not have the normal rows of seats, just a thick padded bench along either wall. It stopped what had to be only about a third of the way down the jet. She wondered what the rest of the space held. Dorsey dumped his bag on a bench a few windows down, then sat heavily beside it. Mike did the same along opposite wall. She stood between, still holding her knapsack tight.

"Destination?" asked Mike.

"D.C." said Dorsey. "This was the best I could do. Space-hops were too noticeable."

"Can I ask what's going on now?" she said.

Dorsey and Mike stared at her, in a way that made her feel strange—like they had forgotten she was there, or could speak, or something similar.

"You've got what you need from Hawaii, right?" said Dorsey.

Mike shrugged. "We think so."

Dorsey frowned. "I mean, there can't be anything else; we followed the clues to a safe and found a code. Now you two just need to crack it."

She supposed she did know what was going on; Dorsey had spewed the relevant information at the hotel. The other parties were on their way back, so they should flee with the prize. She still felt like something was missing, but did not know what question to ask.

Static sounded, then a cheerful voice said, "Any non-existent passengers should buckle up. We'll be taxiing soon."

"Non-existent passengers?" she repeated.

"We're not officially here," said Dorsey, as if that explained it. She supposed it did.

Mike and Dorsey felt along the wall and pulled out paired straps which resolved into harnesses. She found her own. It pressed tightly across her waist and chest, pulling her securely into the padded wall behind. The benches were unusually wide, but she and Dorsey were able to sit with their feet on the floor. Mike pulled his feet in front of him and sat cross-legged.

"Once we hit cruising, we can get back to your genome," said Mike. Carolyn grinned. She looked forward to it.

*   *   *

Carolyn sat with the student's elly-book perched on her bent knees. She scanned through the file Mike had sent her, but he was right. It was nonsense. Occasionally, something looked like it might make sense—scattered collections of characters that looked like DNA sequences or calculations. But it was unreadable. Just like the one he had passed on shortly after they reached cruising altitude. They had tried ordering codons using first the numbers of corresponding tRNAs in her parents genomes, and this was now according to hers. Those were the easier elements to extract; she had gotten him those even before he was able to finish extracting the codon frequency from the genome he had. So she had spent the time Mike's computer scanned for ... whatever it scanned for ... to calculate the actual frequency of codons in her parents' and her genomes. Like the tRNAs, the ordering produced by codon frequency was identical whether or not she used her mother's or father's genome, but different for hers. Clearly they had modified things, her code-loving parents. She wished what code they may have used was more obvious. Yet, obvious codes weren't very *good* codes.

She said, "You said their codes were simple substitutions—they wouldn't have done some kind of double substitution, like this is another code?"

"My algorithms would have deconvoluted that as well." Mike stretched his legs from where they had been folded cross-legged under his laptop. "So nothing there?" She nodded. "I'm almost done with trying your parents' frequency of codons—you can look at that next." His computer gave a low ding, and he pulled it back to his lap. He opened a file and scrolled down. "Hey, I think ..." He sighed. "No, it looks like nonsense again."

She peered over his shoulder. The screen was a solid block of characters, just like the file he already sent.

... XNBF05UL10MMDNTPS2ULMGCL25ULBSA05ULYAC1A188GF05 ULYAC1A188GR2ULTMP0125ULTAQ12875ULH2O ...

It seemed unreadable, but there were elements that recalled biology. DNTPS could be dNTPs, or deoxynucleoside triphosphates, the raw building blocks of DNAs. BSA was the abbreviation for bovine serum albumin, a common stabiliser for PCR reactions, the process used to build DNA … with dNTPs … Her heart leapt with excitement. Similar patterns had popped out of the other decodings, but never quite so close together. Nor so thematically linked. There was Taq, the enzyme used in PCR. And $MgCl_2$, or magnesium chloride, the co-factor for Taq, which enabled the enzyme to work. Of course, H_2O was just water. The only thing she couldn't figure out was the repeated pattern of YAC1A188G; YAC could stand for yeast artificial chromosome, but the rest of it made no sense. Her breath caught. The F right after could stand for 'forward', matching with the R on the repeat, for 'reverse'. PCR reactions had forward and reverse primers. She was sure. "No! It's not! That's a PCR reaction! Look." She grabbed her elly-book, scrolled his screen up just a bit, and sketched:

25 UL 10X RXN BF = 2.5 µl 10X reaction buffer
05 UL 10MM DNTPS = 0.5 µl 10 mM dNTPs
2 UL MGCL2 = 2 µl $MgCl_2$
5 UL BSA = 5 µl BSA
05 UL YAC1A188GF = 0.5 µl YAC1A188G forward
05 UL YAC1A188GR = 0.5 µl YAC1A188G reverse
2 UL TMP = 2 µl template
0125 UL TAQ = 0.125 µl Taq
12875 UL H2O = 12.875 µl H_2O

He squinted, still looking puzzled. "Does that mean something to you?"

"Yes, it's a pretty standard recipe for a twenty-five microlitre reaction." Her eyes scanned down and *sense* popped out at her. She pointed to later in the file. "Here is the PCR cycle, with temperatures and times and starting cycle and everything!" She clapped her hands together. "This is it! We've done it!"

Chapter Forty-One

Cracking the code wasn't the complete breakthrough Carolyn had first thought. They still needed to turn the tight sets of letters and numbers into something sensible, then to understand *why* this had been squirrelled away in the first place. Mike's computer helped with the former task. Carolyn provided examples of translating PCR reactions, gels, and other biological tests, and the computer did its best to insert spaces and decimal places in the correct places. But it still required human curating. She was able to train Mike and

Dorsey easier than the computer, and soon they were all pouring over its sug-gested output, making small adjustments or translating blocks the code-breaking software had given up on.

"We could still be off by some factors of ten," she said, having just inserted decimal places into a series of numbers. "Why couldn't they have used just one more character?"

"What they did was amazing as it was," Mike said. "One more character would have limited the amount of information they could get in significantly."

"What did they do, actually? Do you know the code?" she asked.

Mike pulled his laptop closer to him, almost as if he was going to demon-strate. But then he simply spoke, animated. "It was another substitution code, but on several layers. There are four sets of degenerate codons, those present in two versions, three, four, and six. They made four overlapping codes across your whole genome, each one using a new set of, well, codons—two to five unit sets, depending on how many different levels there were—using the ordered frequency of the codons." His face took on a look of wonder. "Maybe they gave you an extra X for more *writing space*."

"The X is one of the larger chromosomes," she said. "These overlapping codes, I'm not sure I understand."

"You go through your genome four full times, each time reading just one of the sets. They must have done this entirely on their own, as they stuck with the simplicity: the first trip through is the set of codons present in six versions, and they used sets of two these, to get six-squared—thirty-six—distinct char-acters. The twenty-six letters of the alphabet plus ten digits. Then they go through those in four, three, and finally two versions. That last one's the only one they didn't manage to find a way to code out thirty-six items: they used five-set codons, making two-to-the-fifth, or thirty-two, items. They appeared to have left out some digits; when we get to the end I'm guessing we may find some letters doing double-duty."

She took a breath, planning to ask more.

"Perhaps we should keep going and *get* to that end?" said Dorsey, quickly, as if heading off what could have been a long digression. She was disappointed; Mike was going into teacher mode on his own material, and she found it interesting. But Dorsey was right. Figuring out what her parents had said was more important right now than Carolyn understanding *how* her parents had said it. Dorsey said, "I don't understand the point of all of this anyway. Is it just another lab book?"

"I don't know." She thought through what they had seen so far. It seemed to be. "I expected more." She expected an *answer*, not more, endless, mystery.

"It's not really a lab book. It's more like …" She hummed to herself, trying to capture her thought. "… the methods section of a paper. Remember how the lab books had dates, and experiments repeated, and so on?" Mike nodded, even though Dorsey had asked the question. "This is more like instructions." She tilted her head. That wasn't quite right. "Or, maybe like the raw material for a paper. Because there are results, too." Just no introduction or discussion to explain what one was looking at.

"I think we've got the idea of what we're doing down," said Dorsey, with perhaps a bit of exasperation in his voice. "Instead of checking the output, why don't you start reading from the beginning? You'll be the only one who can understand the science."

She hesitated. It was a good idea. Yet … she had liked doing something that made sense. Or was she afraid she wouldn't understand the science if she tried? She swallowed. "Okay."

*   *   *

She had nearly caught up to Mike and Dorsey's translations. What started out as seemingly scattered experiments revealed a theme: mitochondria. She scanned what she had just read one more time. She scrolled down, fingers tingling, already anticipating what she'd see. Yes! Mitochondrial sequences, this time matching the sequence that had been the target of the most recent PCR reaction. Previously, the pattern had been reversed—it was the mitochondrial sequences that were being PCR'd until this point. Now the PCR sequences were appearing in the mitochondria. "It's the Boltzer process!" she said. "They invented the Boltzer Process!" She squeezed her eyes shut, trying to remember the date of the initial Boltzer process paper. "At least five years before Ethan did." Wait, no, it would be longer. Because while this had been hidden away at their deaths, it had been encoded into Carolyn three years previous, and the research that had been hidden away was years before that. "Or a dozen."

"Could that be it?" asked Mike. "The Schwarz Final Findings?"

Her body felt heavy in disappointment. Was all this something that had been known for decades? Yet, there was all that other stuff. "It's not only the Boltzer process. I don't understand all this …" She waved her hand off the top of her elly-book, indicating the material that came before. "They were doing something with the mitochondria and the electron transport chain." The same thing that had formed the start of their research in Paris. There *had* to be more to it.

"Can you figure that out?" asked Dorsey.

"I'd rather keep reading. Do you have more?" She wanted to see the whole picture. If it was really like a paper, minus the explanatory bits, she'd have a better chance of understanding once she read it all.

* * *

That had been the last of the mitochondria. The topic suddenly changed, to artificial chromosomes and their telomeres. She remembered Ethan saying something about that, how it was the other area her parents had hinted about to him. This was about more than just the Boltzer process, she was sure. The cells moved out of yeast, into mammalian cell lines. Then there were animals: mice, rats, rabbits, monkeys. Were they working on something medical? Then, suddenly, everything was in fruit flies.

She chewed her lip. Why move from monkeys to flies? It didn't make sense. But flies it remained, and the mitochondria came back. Her parents repeated the Boltzer process on flies whose telomeres they had modified, and a slew of behavioural experiments followed. But it was a long series of null results: the experimental condition was always not significantly different from the unmodified flies.

Null results weren't news. They were what scientists despaired of, meaning their hypotheses were not supported. Not even strong enough to disprove a hypothesis, null results meant 'we can't say'. Why would this be recorded, at such great expense? At least since the animals had appeared, the experiments were also dated, and she was able to place the fly work to a full five years before her conception ... creation? Meaning the Boltzer process had been found by the Schwarzes before Ethan even moved into mitochondria.

"We should get some shut-eye," said Dorsey.

She joined Mike in glaring at him. She didn't want to stop. She wanted to keep reading; she wanted to *understand*.

Dorsey stood. "We all need to eat and sleep. Fuzzy brains won't decode well. I'm sure fuzzy brains won't science well, either."

She nodded reluctantly. He was right. Dorsey opened a door in the bulkhead ahead of them and pulled out a rack of trays.

"What's there?" asked Mike.

"Fish or lasagne."

"Fish," said Mike.

"I'll have lasagne," said Carolyn.

Dorsey pulled out three trays and vanished around the corner. She had not even explored their small passenger area, so intent had she been on decoding her genome. But clearly Dorsey had, as he reappeared carrying two trays with steam rising.

She kept reading as she ate, but her eyelids grew heavy. The screen of her elly-book swayed; no, she was swaying.

"Here." Mike tossed her a blanket and pillow from an overhead compartment. "Make yourself comfortable."

She slid the elly-book and her now-empty tray down the bench and stretched out, placing her pillow just in front of it. She pulled the blanket over her body, which was now feeling quite chilled. Some sleep was a good idea, after all. She let her eyes drift closed, visions of fly experiments dancing in her head. The fly designations had gotten longer and longer, reaching three and four digit codes. Or maybe they were an age? No, flies didn't live that long.

She tensed with excitement; her eyes flashed open. How long did flies live, anyway? A month or two, she thought. But those protocols …

She sat up and pulled the elly-book towards her. Yes, some of the protocols included training the flies—the experimental group, only—for at least three months. Flies shouldn't live that long. She scrolled up. She had assumed each experiment was a new group of flies, but she realised that the description of *making* the experimental group existed only once. She had assumed because, unlike a lab book but like a paper's methods section, they did not need to repeat what was already known. But was there even anything to repeat?

She scanned the material she had looked at before, getting more and more sure. These were the same flies. The 'designations' she had thought about, on the edge of sleep, were their age. In days. Some of those flies were years old.

She scrolled up further, looking at the mammal experiments. But mammals lived years anyway. Yet … one line of mice was a short-lived mutant. The experiment hadn't lasted long enough, though; while none of the experimental group had died, not enough of the controls had either to make any significant difference. Her parents had moved to flies for their short lifespan.

And the null results … The *years* worth of null results. They were checking to see if the flies behaved differently than unmodified flies. Mitochondria and telomeres. The two axes of aging: as organisms aged, their mitochondria got more inefficient and produced more free radicals, damaging DNA and aging cells; meanwhile, telomeres shortened, leading to still-yet not understood changes in gene expression associated with aging.

Her parents had modified the mitochondria, through the Boltzer process, and the telomeres of flies, which then lived for many, many times their natural lifespan. Immortal flies.

She raised her head, looking at her sleeping companions. She knew the Schwarz Final Findings. Her parents had discovered the secret to immortality.

Chapter Forty-Two

She couldn't sleep now. She scrolled back to the start of her translation, determined to understand how they had done it. How had her parents made immortal flies?

She inserted her own notes, a sketch of what could have been the missing discussion to go alongside these methods and results. Now the sequencing of mitochondrial genomes made sense—they were measuring the genetic sequence that produced a variety of outcomes. But how did they know to use the Boltzer process on just that one sequence? She reminded herself that this was like a paper, not the lab books she had been reading. Only the one successful result would be included, not the null results—at least not for this part! She grinned to herself, reading more, sliding the experiments into her own knowledge of the future of mitochondrial research.

She chewed her lip. Again, her parents were uncomfortably close to her own research. Not only the electron transport chain, but dynamics of mitochondrial turn-over were key features in the experiments described. Humour—or was it hysteria?—bubbled. Scientific research paths definitely couldn't be passed down through the genes; it was a just a great coincidence. She was far older than her parents had been when they discovered immortality. Clearly she wasn't just following in their footsteps.

Her thoughts were getting more and more ridiculous. She shut down the elly-book and pushed it back up the bench. Her eyes closed before her head hit the pillow.

*   *   *

She woke, filled with excitement. Had it been a dream? She pulled the elly-book towards her. No, it was just as she remembered from her sleep-fogged state; her annotations were now interspersed in the translation. Mike was stirring as well. Dorsey wasn't visible; then he appeared from the front area, carrying trays.

"You're up," Dorsey said. "Timed it well. There was just one breakfast choice."

She accepted the tray and dug into an ambiguous yellowish item that smelled vaguely of egg and cheese. It tasted of nothing, but she was ravenous and ate eagerly.

"How much longer?" Mike asked.

"A few hours yet," said Dorsey. Mike groaned. "I know," Dorsey said, sounding defensive. "But we needed to be inconspicuous. These little planes aren't used for important things, because they're too slow."

"At least it gives us time to keep decoding," said Mike.

She finished her breakfast. She should say something about what she'd found. Suddenly her conclusions seemed silly. Perhaps she had read more into it than was there. *Immortality.* It had seemed an obvious answer in her sleep-deprived state. Her morning self was embarrassed by the hyperbole. But she needed to say *something*; she did not want to repeat her embarrassing reticence from the very beginning of this adventure, hiding key pieces of information. "I think I've figured it out." Mike and Dorsey looked at her; Mike half-rose. "I mean, I have an idea. Maybe. I've got some clue."

"What is it?" Mike said, eager.

"I think they found the secret to … longevity." She replaced the term at the last moment. Immortality was too magical, anyway. The flies were extraordinarily long lived, not unkillable.

"How so?" asked Mike.

She pulled the elly-book to her. "These experiments here are about mitochondria." She scrolled down. "And these are about telomeres." She scrolled to as far as she had gotten. "They genetically modified these flies, based on this research, and they lived *years*."

"Is that long?" asked Dorsey.

"A normal fly lifespan is about 28 days," she said. "It's incredibly long for a fly."

Mike whistled. "But it's just in flies? That's … well, just flies."

"They did mammals too," she said. "Even monkeys. The techniques worked. They must have gone to flies for the effects—even mice normally live a few years, a decade for a well-looked after lab mouse. It would take far too long to see if there was an effect on lifespan."

"So do you think this could be it?" asked Dorsey. "Longevity? Physicians have been trying to understanding aging for a long time. Why would you hide away such a breakthrough?"

"Corporate," said Mike.

"Ethan said they didn't think the corporations should have it," she said. "They wanted it to come out in the public domain. But they died … were murdered … before it could." Her chest filled with a strange pride. Her parents weren't the deceitful crooks she had thought most of her life. They were so honest they had died rather than let the secret of immortality fall into corporate hands.

"Imagine corporations owning longevity treatment," said Mike. "It'd cost a fortune. Only the rich could afford it, then they'd stay alive and in charge and rich."

"They're already in charge and rich," said Dorsey.

"But they'll die," said Mike.

"Their children will inherit. I don't think it'd make a big difference."

She agreed with Mike. It was within living memory that corporations had gotten a strangle-hold on the world. Anticorporate radical communities existed, tolerated for the most part. But imagine if companies could make sure their employees lived forever? Stratification in society would increase by orders of magnitude. There might be a way back now. With immortality in the hands of corporations only, there was none.

"We're not done decoding yet," said Mike. "Let's see what else is there, and if this holds up."

She kept going from where she had stopped adding notes. By the time she got back to the fly material, Mike and Dorsey sent more translated bits on. A large amount of the mammalian and fly work was experiments showing no harmful effects, other than what appeared to be a slightly increased metabolism. Then, the study subjects changed again. To humans.

Chills ran down her spine. Who had they done this to? Were there immortal people out there already, employed by Vivcor? Reading further, she saw there were just two subjects, one male and one female. Toxicity screens would normally have far more people involved. A sudden idea struck her. She pulled up her parents' genomes. Had they been maverick enough to test it on themselves?

She copied the relevant sequences from the telomeres, plus the few promoter regions of telomerase and telomerase regulators that were involved in that modification, and searched. She stared at the perfect red matches. Yes, they had modified themselves. Queasiness struck her stomach. If she was a near-clone of her mother … She searched her genome. It was the same red perfect match.

She had the immortality treatment as well.

Chapter Forty-Three

But it wasn't just the telomeres. The mitochondrial treatment was required too. She turned back to the translation she was annotating. A little further on, and yes, they had modified their own mitochondria as well. Would they have modified hers? She had no way to tell, not without a lab.

The mitocyls had confirmed she had her mother's mitochondria. According to Mike, the sequence had been stored when her mother was a student in the US. That would have been an unmodified sequence, which she had matched. But she didn't know what the mitocyl tested; it could be just a few key SNPs. They had the mitocyls, though. She could find out.

She kept going through the translation. They had tested themselves, physical and mental tasks, and seemed to be normal. She had at least some of the modifications, and *she* seemed to be normal. Or at least lacking any obvious health issues due to having an unusual set of telomeres.

She reached the end. "That's it," she said. "It's a longevity treatment. They tested it on themselves." And me. "Didn't help them live very long, did it?"

"No," said Dorsey. "It seems such a silly thing to hide and die over."

She shrugged, not wanting to argue over whether the corporations could have been responsible with the longevity treatment. "Will you still be trying to solve their murder?"

"Of course!" said Dorsey. "We've got a lot more leads now."

She scrolled back up through the elly-book. The Schwarz Final Findings. How long ago had it been since that Sandslin rep had jokingly mentioned them? Had he been fishing for information? She wondered if the corporations had any idea what it might be, or if they had been as in the dark as she. "What about me?" Her life was destroyed.

Mike leaned forward and patted her knee. "We'll get you and Ethan new identities. You can go back to doing science."

"But what about …" She trailed off, suddenly realising that along with the secret of the elly-book's tunnelling capabilities, there were more things she had kept from Mike and Dorsey. Like the existence of her daughter Ellen. Sandslin had not known, and if they had not, the government agents following them probably didn't either. For some reason, she didn't want to rectify that right now.

"Or whatever you want," said Mike.

"What about Susan?" she asked, trying to make her previous words make sense.

"She can get a new identity too," said Dorsey. "We could even use her skills over here, if she wanted." *The United States will take care of you*, Carolyn heard in his undertones. She smiled slightly, trying to act comforted. But she wasn't. She wasn't sure she trusted anyone but the anticorporate crowd. Or maybe just Susan. Mike looked at her, and she thought she saw understanding in his eyes.

*   *   *

They landed in the rain. She followed Mike down the airplane stairs, Dorsey behind her. The damp air was cold, a stark contrast to balmy Hawaii; the raindrops hitting her head felt like needles of ice. She squinted at the sky: it was dim and grey. She could not tell if it was morning, midday, or evening. Her internal clock was spun about. They walked quickly across the tarmac until they were under a covered walkway.

"I need to report in," said Dorsey.

"I can take Carolyn to a safe house," said Mike.

"Use of one the Bureau's," said Dorsey. "We need her."

Mike nodded. But she frowned. Why did they need her? Surely she could not shed any more light on her parents' murderers. She followed Mike, who walked quickly, keeping pace with Dorsey until they entered the building. They wound through featureless wide corridors until they emerged at the edge of a large parking lot. Mike strode confidently down one aisle, while Dorsey turned off to the left. She ducked her head from the icy rain and trotted to stay with Mike.

He opened the door of a small grey car, old, with signs of salt damage around the wheels. He grinned at her. "Expecting something fancier?"

"Well, yeah." Didn't spies drive sports cars?

"I'm not James Bond," he said. "I get a government salary, and all I care about in cars is whether or not it can get me there."

She sat in the front seat and looked at the all-manual dashboard. "Or rather, whether you can get it there," she said with a smile.

"Got me," he said. "I'm not having autodrive in my personal vehicle. *That's* too dangerous for a spy."

But autodrive was meant to be safer than humans. Although maybe he meant someone else taking it over. But it was also meant to be unhackable. *Like docfilm*, she thought, remembering Susan's modification of their train tickets; like elly-books were meant to be isolated from the net. Not everything was as it seemed.

He reversed out of the parking space and drove from the airport. The road was huge and wide, far different from the small windy roads in Hawaii or the tight streets of London and Europe. Train tracks ran down the median, separated from the road by a wire fence; the road itself had several separated sections—they were in the most central set. "Where are we going?"

"I know a Bureau safehouse in Reston," he said. "It's not far from the airport." True to his words, they soon exited—first onto progressively more outer sections of the road and then onto more normal-sized streets. They drove past a quickly flashing array of shiny high-rises, mid-sized buildings patterned with sculpted designs, and low, wooded communities. It reminded her of the anachronistic centre of Bucharest, except the suburban version.

He pulled into a parking garage connected to one of the mid-sized buildings and parked several levels up. "Follow me. Don't say much until we get to the room."

"Okay." She was nervous. She no longer understood what she was doing. Before, she had been on a mission to find her parents' secret. Now she had it. What was the next step? Why was she still hiding?

She tried to remember her initial determination: she would destroy the secret, or release it to the world. She had done neither. Mike knocked a pattern on a door in a carpeted corridor, then whispered some low words into a speaker by the doorframe.

The door opened. Two … agents … stood inside. Unlike Mike and Dorsey, they looked the part. They wore tailored suits and bright white shirts. "Hafal," Mike said when one of the agents shut the door behind him. "Dorsey should have sent—"

"He did. Is this her?"

"Yes." Mike gestured. "This Carolyn. Carolyn, meet …"

"Agents Smith and Jones," said the taller of the two men.

Carolyn's eyebrows rose. *Really?* "Who's who?" she surprised herself by asking.

"I'm Smith," said the taller man.

"I'm Jones," said the other.

"This way." Smith led her from the foyer into a small apartment. There was a kitchen connected via an open counter to a large living space with a TV, several chairs, and a sofa. Behind the sofa stood the makings of small gym: an exercise bike, treadmill, and what looked like it might be a skiing machine. The television sat in front of a pair of large windows, covered with translucent drapes. Smith gestured at the windows. "We're six stories up. The window doesn't reveal the inside, but don't get closer than the TV, and definitely don't move the curtains." He walked to the left, through a door to a bedroom that looked more like a hotel room, complete with mini-fridge, wall-mounted ironing board, its own TV, desk, and small table with chairs. A door on the interior wall opened into a tiled room: likely the bathroom. "Here's where you'll stay. We've got the same on the other side. You can come out to the main room and kitchen as you wish, for now, or stay in here. We'll let you know if you need to stay here at some point." Smith pointed to a black grill on the wall. "This is the intercom. If you need someone, press the grey button and speak. We can't hear you unless you press it." She was not sure she believed that. He smiled, but it didn't reach his eyes. "Don't worry. You're in good hands."

Smith turned and left Carolyn alone in the room with Mike. She put her bag down on the bed and crossed her arms. "Can we talk here?"

Mike looked at the grill suspiciously. He must have had the same thought as she. "There's not much to discuss. Dorsey or I will come get you."

"You're leaving?" It came out too loud, fast, and panicked. Of course he was leaving. He had agent-things to do.

"I need to check in and see what's happened to my reports." He frowned. "And what other teams might be on the case." He probably meant the colleagues who had showed up in Hawaii, whom they had fled. He was being circumspect. He must believe the room was under surveillance.

"But what—"

"We can cover things later," he said, a strong emphasis on the final word. He definitely thought they were being overheard. He looked back through the open door. He put his hand into his pocket and came out with his phone. He fiddled with it briefly, then handed it to her. "I've got another in the car, you can have this." He went to contacts and scrolled through, stopping at 'Too, Me' and patting it. That would be his other number.

"Do you think—"

"Don't worry, they'll feed you lunch in a bit." He patted around his jacket and handed her a packet of crackers. She recognised the wrapper from dinner in the plane. "But this will tide you over." He guided her hand with the phone to her trouser pocket.

Fear seemed to stick her feet to the floor and her tongue to the roof of her mouth. At least it kept her from making some inane comment. She did not like him acting so cautious. Why was he leaving her here if he thought there was some danger?

She remembered Jenny-heart from his phone. Why would she expect Mike to be attached to her in any way? They had found the secret; he had everything he needed from her. Both Mike and Dorsey had her decoded genome. Although only she had the annotations she had added to her elly-book.

He grinned disarmingly. "Don't worry. Dorsey or I will be back in no time. Before dinner. Or before you sleep, for sure."

"For sure," she repeated. He gave her a quick hug, and she resisted hanging onto him like a life preserver. She even gave a little wave. She placed her hand on her pocket and felt his phone. She hoped he really came back.

Chapter Forty-Four

Neither Mike nor Dorsey returned by night time. Smith and Jones were unfailingly polite, but after an awkward silent lunch in the living space, she retreated to her room with dinner. They signalled night-time by coming to collect her plates, wishing her a good sleep, and turning out the lights in the kitchen and living room.

She sat on her bed with the door closed, stiff, afraid to change lest she be watched somehow. Yet Mike had handed her a phone and made movements freely; he must have thought any surveillance was audio only. She pulled out the phone and looked at it. *Audio only!* She could text. She went to contacts and scrolled through. Her heart jumped again at 'Beane, Jenny ♥'. Mike was entitled to his own life, she reminded herself. Such adventure as the last week was par for the course for him. He could not fall in love with each female he encountered in the course of his work.

Shock at her own thoughts filled her. Her body chilled, and she froze in place, hand still on the screen. Where had that thought come from? She had not fallen in love with him. Absolutely not. She bit her lip and kept scrolling. She found 'Too, Me'. The humour of the entry made her smile. So much about Mike made her smile. She had not fallen in love. But she could not deny she enjoyed his company. That was okay, though. One could enjoy a friend's company.

She probably wasn't even a friend, either. Just another day on the job. That was her. She gripped the phone tighter, its smooth corners pressing into her palm, and concentrated on her text message: *Where are you? It's bedtime!*

The thought that that might be misconstrued by a Jenny-heart passed her mind. She was fine with it. Mike had stood her up. He said he or Dorsey would be back by now.

She stared at the phone, willing it to reply, but it didn't. There was someone else she could check in with. What time was it in Paris now? Probably the middle of the night, she calculated. Susan would think she was still in Hawaii, and they had set up midnight as contact time. She squeezed her eyes shut tight, trying to remember the time difference between Hawaii and the east coast of the US. It was something like five or six hours. She set an alarm on the elly-book for five a.m.

She lay back. There was nothing else but to try to sleep.

*   *   *

The elly-book woke her, but the first thing she checked was Mike's phone. Finally! There was a reply: *Sorry, held up. Will see you tomorrow.*

Not extensive, but it did have an apology. And another promise. He had better show up later that day. She opened the tunnel app and typed in Susan's code. Relief filled her as the screen swirled away and the familiar split-view appeared.

WHERE HAVE YOU BEEN!!!!! appeared almost immediately in the text bubbles.

Sorry, Carolyn typed, suddenly feeling for Mike. *Too much was happening. There was no chance.* It was not entirely true. There was that one night, before she realised that her genome was the code, when she could have tunnelled Susan. But then Susan would have spent the intervening days thinking all was lost. *We did it! We found my parents' message.*

What was it?

This was harder to type than she had imagined. Again, suddenly, the 'Final Findings' seemed ridiculous. She remembered Dorsey's words. *Such a silly thing to hide and die over.* Was it? She took a breath and typed: *They discovered a longevity treatment. My genome was the code.* Wait, Susan wouldn't know Carolyn really was synthesised. *Oh, I'm not a hoax!!! I am synthesised. They hid what amounts to a whole paper inside me, in my genome.*

The screen showed a long line of Carolyn's text bubbles. She halted, nervous. What was Susan making of this?

WHAT???? finally appeared. That didn't answer Carolyn's worries. Then, *That's incredible. You DID have it, after all! Longevity?*

Queasiness in her stomach made her realise she cared a lot about what Susan thought of this. Would she be dismissive like Dorsey, or understand, like Mike?

It seemed ages before another bubble appeared. *I can see why they hid it. Can you imagine corporations with exclusive hold on long lives?*

She relaxed. Susan understood. Another person, who wasn't a supposed-to-be-paranoid spy saw the same dangers she had. *I know. But now we can let everyone know. Publish their final paper.*

What if Vivcor claims rights on it and patents it?

She had not thought of that. The results did belong to Vivcor. Vivcor, the corporation she had always thought of as her saviour. Yet was it them who were responsible for her parents' deaths? She still didn't want to believe it. She remembered the Vivcor interns, the coordinating woman, those old men. They were people, not a faceless corporation. But that was the problem, wasn't it? That was why the corporations had been able to take over the world, because there was no 'face' anyone could identify and call the 'baddie'. Individual people working for pay—like her parents—and maybe some people at the top in charge of boards or things like that that made decisions about 'units' and 'stock' and had lost the human connection to those they made decisions about. The corporations had grown to be greater than the sum of their parts, and individual people were like a member of a termite colony, suddenly wondering where that giant mound had come from.

They had all contributed, even Carolyn. Especially Carolyn, doing research that the corporations also owned, even if she had free disclosure agreements. But what about her parents' research? Should she destroy it, instead?

I don't know. Her throat caught at the thought of losing her parents' final effort. What they had died to hide. *Maybe it should be destroyed.*

You can't destroy yourself! Susan wrote.

Her body tingled, as if she were feeling every one of her cells. Susan was right. It was not possible to destroy her parents' secret. She didn't know what other records might exist; someone else must know she was truly synthesised. The letter writer? Patty? Not Ethan.

Her previous wonderings about who in her parents' lab had been complicit in the hoax turned into wondering who knew the truth. But if the hoax remained the story, they would all think they were duped. Her brain spun trying to follow all these strands at once. What if she had the full treatment and didn't age? *That* would become obvious.

You're right, she typed. The treatment wasn't complicated. Or even expensive. It was just the knowledge that was hidden. Vivcor might *try* to claim it, but they couldn't claim the Boltzer process. They couldn't claim the gene manipulation, now standard techniques—she didn't know enough about that area to determine whether her parents had been ahead of the curve there as well, but perhaps she could guess they were. They had done so much ahead of their time: synthesising her, discovering the Boltzer process, … finding immortality.

A lot of it has come out already, she typed. *They described the Boltzer process ages before Ethan did!* She grew calmer as she wrote, reassuring herself and Susan at the same time. *The techniques are simple and cheap. Even if Vivcor did try to patent some of it, anyone could do it in their basement.*

Their basement? typed Susan.

A well-equipped basement. Carolyn smiled. But it had not been long since the age of DIY-biology, when people really did genetic manipulations in their basements and garages. Maybe they still did.

What are you doing now? Are you coming back?

What *was* she doing now? She wasn't sure. *I don't know. Dorsey seems to want me for something more for his investigation. Mike said they could get us new identities, after.* She bit her lip. *They don't seem to know about Ellen. Have you heard anything through your people there, from Bae?*

Yes, she's fine, typed Susan. Relief flooded Carolyn, along with a desperate longing. She wanted to be home with her daughter. But there was no home anymore. *Bae says she's upset you didn't take her with you.*

A thirteen-year-old on this trip? Carolyn would have been a complete mess worrying about Ellen's safety. But it was so like Ellen to want to be involved. *You can tell her that I expect to be back soon.* She would. She would get home, somehow.

Chapter Forty-Five

Carolyn woke with her face on the elly-book again, to the sound of knocking on her door and Mike's voice. Her heart sped with joy. He really did come back! She had fallen asleep after an indeterminate time chatting with Susan, sharing as much as she could of her last hectic days: Mike's dubious colleagues, the other company, her parents' music-based message, the people at C.A.V.E., Mike's anticorporate sister, the whole thought-it-was-nothing-and-realised-it-wasn't thing, and their final long flight back with all the decoding.

She blinked sleep from her eyes. "Yes?" she called.

"Can I come in? Don't have much time here."

"Just a moment." She rushed to the bathroom and changed back into yesterday's clothes; she needed to do some laundry. She used the toilet then emerged, saying, "Come on in."

Mike opened the door and slid through. He closed it behind him. "Dorsey's people have gathered some more science-types for you to talk with. I've got the code, but only you have the translated versions from both me and Dorsey. Would you mind passing me a copy?"

"Um …" She didn't mind, but she was unsure what was the connection between his first statement and the question.

"I've got some things to deal with, and I'm not sure when I'll get back to you. I'd really like to make sure there are more copies of the information."

Her speeding heart moved from excitement to worry. "Am I in danger?" He wasn't suggesting that the FBI might try to harm her or steal the secret, was he?

"No, no. Not from the Bureau. Not you." He opened his laptop on the small table. "I'm not so sure about your parents' research, though. There may be some that aren't happy about the idea of releasing all of this. I'd hate there to be some kind of 'accident' with the only copy."

"But *I'm* a copy of it. Just encoded." She handed him the elly-book as she spoke, filled with his paranoia. He was speaking freely; he must not think they were being listened to right now.

He looked up from plugging in the elly-book. "You are, aren't you?" He shook his head as if shaking off a thought. "So it can't ever be lost."

Unless someone killed her, she thought. Would that have been her first thought two weeks ago? Probably not. She had lived in a different world then. A nicer world, with magnanimous corporations and honest police. How did she know the FBI wasn't secretly owned by someone, despite what Dorsey had claimed? But she nodded as if reassured. Mike surely would know if his government was compromised, wouldn't he? She remembered his dodgy colleagues. "Could I maybe have the decoding … whatever it might be … in return? The pattern, or something?"

"I'll copy over the script my programme created." He tapped at his keyboard. "If I get a chance, I can write up the pattern for you, too. It is reasonably simple; as I said just substitution." He stilled his hands which had begun to dance. "But you can see it in the script." He closed his laptop, unhooked the elly-book, and stood. "Thanks." He moved towards the door. "You've got my number. Call if you need me."

"Is this goodbye?" she asked, panic rising.

"I'm sure it's not." He shifted his weight from foot to foot. "I'll help you and Ethan get sorted in your new lives."

"And Susan?"

"And Susan," he confirmed. He stepped forward, fast. He put a hand on her elbow, and when she didn't flinch away, put the other around her back in a hug. "I *will* see you again. Even if it's not my job. Promise."

She smiled. She believed that more. "I'm glad." She wanted to say more, to chat, to get him to describe her code some more, or anything that got him in that eager, animated mode. But she remembered *Beane, Jenny-heart*. She returned his embrace, lightly. "See you soon."

He vanished out the door, and she sat heavily on the bed. She was hungry. She opened the door and leaned out. The two agents were sitting on the sofa, facing the blank TV. "Smith?" she called out.

The shorter man turned. "Yes?"

Carolyn had thought he was Jones. "Breakfast?"

Smith or Jones or whatever his real name was moved to the kitchen. "I'll show you."

She made herself some cereal. She had just finished when Dorsey entered the apartment. "We've got some experts for you to meet," he said. "Come along."

"Let me get my stuff." She stood.

"You don't need anything," he said.

"The Final Findings?" Her voice rose in disbelief.

"Well, yes, I suppose those. But we already have a copy."

When had he gotten it? She tried to remember the final part of the plane flight, but couldn't identify a time she had specifically given the elly-book to Dorsey. She couldn't remember that she kept it from him, either. "I'll bring my own, if that's okay." She moved back into her room, not waiting for an answer. She remembered what Mike had said about an 'accident'. Leaving the elly-book unattended with Smith and Jones was ripe for a such a situation.

Most of her bag was still packed from the flight; she had pulled out only her pyjamas, toiletries, and the student's elly-book. She had already been nervous, and Mike had made her more so. She pulled out her dirtiest pair of clothing, those that had hiked to and from a cave and spent the night at the C.A.V.E. She shoved her toiletries and the elly-book into the empty space. Leaving some clothes and her pyjamas behind would look less suspicious, she hoped. But she wanted to be mobile. "Ready." She shouldered the knapsack and joined Dorsey again.

*   *   *

They entered a non-descript building in one of the low, wooded communities. She thought she might have glimpsed a gym and a pool behind a fence to the left. She was not sure if it was a hotel or a community centre. Or another government safehouse thing, she supposed. Dorsey stopped at a desk manned by two young men. He placed his mobile phone, laptop, and a docfilm pad on the desk, then took a plastic token from one of the men.

The other man looked at her. "Communication devices, please."

"What?" she asked.

"You need to hand over any communication devices to enter," the man said.

"It's okay," said Dorsey. "Just procedure. You'll get them back." He lifted the plastic token.

"I don't have …" Oh, she had Mike's mobile phone. She slowly took it out. Getting it had felt like a lifeline, and now she had to hand it away? She had not even memorised or copied over Mike's 'Too, Me' phone number, not guessing she would be parted from the phone.

"Place your bag on the scanner, please, and walk through." The man handed her a plastic token and gestured around the corner. She leaned over to see to an airport-like security archway, but nearly twice as thick. Dorsey dumped his shoulder bag on the indicated surface and went through the arch. She followed suit. Clearly more than a hotel or community centre, she decided.

"You've got some docfilm here." The man opened her knapsack and pawed through. He passed the student's elly-book without notice, lifted her parents' ancient elly-book and inspected it curiously, then pulled out her wallet. She must still have the tickets to Stirling. Her mouth grew dry. Maybe even the image of Ethan that Susan had given her way back when. But why was she worried about them seeing that? They knew Ethan; the FBI had contacted him first. But they did not know *she* knew Ethan. Or did they? It was Mike who had not mentioned that to his dubious colleagues.

"Sorry." She plucked the wallet from his hand. "Forgot about these." She pulled the tickets and associated receipts from her wallet. Ethan's photo *was* there; she slid it out too and slipped it into the middle of the other stack. "Here you go." She handed the stack of docfilm to him and put her wallet back in the bag.

"We'll run this through again." He walked back to the start.

Carolyn finally retrieved her belongings, minus Mike's phone and her docfilm, and followed Dorsey down a corridor. They entered a room where two women and three men, all appearing a few decades older than her, sat on grey rolling chairs around a rectangular varnished-wood table, each with elly-books in front of them. The only electronics allowed in this internet-free zone, Carolyn realised. She wanted to clutch her own elly-book. Even the FBI didn't know about the tunnelling app.

"These are our experts," said Dorsey. "Professors …" He spun through their names so quickly that Carolyn failed to catch them. Professors would mean scientists, who worked for Universities. So they were not in government employ. Although perhaps they were retired, like Ethan. "Everyone should have had a chance to read the Schwarz notes along with Carolyn's annotations already. This meeting is to see if you all have anything to add or change to what's there and to find out what you need to go forward." He gestured at an empty chair around the table. Carolyn cautiously stepped forward and sat in it. Dorsey retreated to a chair pushed into the corner of the room.

One of the women angled her elly-book towards Carolyn. It showed the file of Carolyn's annotations to the translated Final Findings. "I wanted to ask about this part. You interpreted this modification to the promoter for the telomerase reverse transcriptase unit to be capable of enhancing its production. I'm not sure I see why."

Carolyn pulled out her own elly-book and scrolled to the same place. She bit her lip, trying to remember her thinking. She had been so tired when she did this, her notes barely made sense. "Ah! You have to look a little further on. Sorry, I should have noted it later instead of back here. The experiments in human cell lines *did* show increased expression, even though the yeast didn't." She found the appropriate place in her file and showed it to the woman.

The woman nodded, and one of the men cleared his throat, tapping at his own elly-book.

*   *   *

By the time someone came in with sandwiches for lunch, it didn't matter that Carolyn had not caught the names of Professor-so-and-so, as she knew them as Jayla, Abby, Bill, Tony, and Gabriel, and their questions revealed their areas of expertise to range from mitochondria, like her, to fly behaviour. The FBI had worked quickly to pull in people with exactly the set of knowledge covered by her parents' material.

The fact that they were in some kind of secure government facility was nearly forgotten as they poured over the science, seeing for the first time research that had been hidden for decades. The consensus grew that Carolyn's interpretation was right. The Schwarzes had found a longevity treatment, showed its performance in cells ranging from yeast to monkeys, proved its life-extending properties in flies, then tested it on themselves. "Unfortunately the human experiments did not last long enough to see if there were any long-term effects," said Gabriel, the fly behaviourist, sounding disappointed.

Carolyn nearly opened her mouth, in protest of the cavalier attitude to the life and death of her *parents*, then halted herself. The reality of her situation crashed down on her. She did not know what Dorsey had said to these scientists about her—why she had been the one to find and annotate the notes. Did they realise she was Baby C? They might think she was just another FBI expert, who happened to be on the ground in Hawaii. In either case, she was not quite ready to reveal that she might very well be the long-term experiment

Gabriel was looking for. Although these were the people who might be able to help analyse her mitochondria to find out.

"We'd have to start all the way back with mice," said Jayla, an expert on human aging. "And run them far longer to get enough evidence to support moving onto Phase I trials."

"Have any of these people,"—Bill waved a hand in a vague circle above his head—"said anything about who's going to fund this research?"

"I'm not sure we'll be the ones doing it," said Gabriel.

"Then why'd they get us together?" asked Bill.

Carolyn looked back at Dorsey in his corner, but he did not appear to be paying attention to the conversation. "What *have* they said to you?"

Jayla said, "It was just—"

Dorsey stood and walked to them, listening to his wrist. "Supper is ready. I know some of you have families to get home to tonight, so I think we can call it a day. If you want a meal, feel free to stay, otherwise I'll see you in the morning."

Jayla, Bill, and Tony gathered their elly-books and left the room, leaving Carolyn, Abby, and Gabriel to follow Dorsey. It had been subtly done, but Carolyn was sure Dorsey had broken up their conversation at that point on purpose. She gripped the straps of her knapsack tightly. She was not sure whom to trust.

*   *   *

Carolyn hugged her knapsack to her chest like a shield. "Just my pyjamas and some clothes, I think," she said, answering Dorsey's query about what she needed from the other safehouse. He raised his eyebrows, but didn't say anything. Did he think it odd she didn't need her toothbrush and other toiletries? "Why can't I go back there?"

"It's better that you're not travelling back and forth," he said.

"What about the investigation? Don't I need to do anything for that?" She didn't know what, but Dorsey had claimed he still needed her.

"Not at the moment. But that's why you need to stay here. People will soon start to realise what's going on, and I want you in here, safe, in case there is any pushback from Sa … any other players."

Here, chatting with their tame scientists. She was one of their tame scientists. She had not missed how he avoided saying the name 'Sandslin'. She could cooperate, for now. "Can I get some laundry done?"

Chapter Forty-Six

She stood at the open window, cold night air gusting under the sleeves of her pyjamas. Beyond the waxy leaves of the rhododendron bushes just outside her window, the suburban community was mostly dark. Only one dim, yellowish streetlight was visible far to the left. Faint light from scattered windows of widely separated houses flickered through leaves waving in the wind.

Too many days had passed. Nothing of significance seemed to be happening. She had failed to get even a walk outside, where she might be given back Mike's phone and send a message to him. She could just crawl out the window, but what would she do with no phone and no idea where she was? They probably had surveillance. She wouldn't get far.

She could not tell if Dorsey and the FBI were isolating her on purpose or it was just a side-effect. But why did they even have a place like this, where no communication was allowed? It made her uneasy.

She refrained from tunnelling Susan, worried that even if the connection was untraceable online, they might have some devices that could detect internet activity. She wished Susan or Mike were here, to advise her on whether that was even possible. But if they were here, she wouldn't be needing to try to figure out how to contact them.

She was unsure what sort of surveillance the place had. They had interrupted Carolyn's and the other scientists conversations often enough at key points, even without anyone but them in the room, that she was sure someone was listening. Were they looking, too? So she made sure to change only in her bathroom and did not even glance at the tunnelling app. She closed the window and perched at the head of the bed, elly-book on her knees. She pulled up the blankets, chilled. She opened her translation of her parents' research and started reading, this time from the beginning.

*   *   *

She woke, bleary eyed, having spent longer than she had intended pouring over her notes, again. She would see Abby and Gabriel at breakfast—the two were staying here as well, while Jayla, Bill, and Tony seemed to live locally. The fact that the three scientists came and went each day added to her confusion. Maybe it was for her safety, after all.

"Actually get some sleep, Carolyn," Abby admonished when she joined her at the cafeteria table with a plate of scrambled eggs.

Carolyn inhaled the strong scent of her coffee. "I keep meaning to."

As always, some suited agent-types sat quickly down beside them. No private chit-chat for the scientists. Carolyn's muscles tensed. If it was for safety, why couldn't they talk in peace?

"How much longer do you think this will take?" Carolyn asked. That had to be a safe subject.

Abbey shrugged. "They told me a few weeks. It's only been a few days yet. What did they say to you?"

One of the agents shifted, as if preparing to make an interruption.

"They didn't give me a time frame," Carolyn said. The agent settled. She seethed internally, tinged with fear. What had she gotten herself into?

* * *

Carolyn leaned forward in her chair. "I don't understand. Why do we need to design enough studies to get to a Phase I trial? Surely we've got the basic idea here. We just write it up understandably. Enough people will jump on the chance to confirm the Schwarz Final Findings that its accuracy will be settled in no time."

The five scientists stared at her like she had suddenly grown wings. "They're not going to release this," Bill said. "That's why we signed all those secrecy papers."

"What? They can't keep this secret!"

"Of course not," said Jayla. "I'm sure they'll trickle it out through the scientific community."

"Us," said Bill.

"Not us," said Gabriel.

"If we come up with a good proposal, it could be us," said Bill. "That's why *I'm* here."

"This research doesn't belong to the FBI or even the United States," said Carolyn.

The door slammed open and Dorsey raced in. "Who owns the research is not your concern."

Carolyn had wondered how the conversation had gotten so far without interruption. But she wasn't going to stand for it this time. She stood and faced him. "It is very much my concern."

"If something was released as the Schwarz Final Findings, it would just be seen as a joke," said Dorsey, patiently. "Their credibly was broken with the Hoax." *Me* seethed Carolyn. But she wasn't a hoax. "We can't risk that. We need it to emerge in the natural process of science."

"So they don't get credit? Are you telling me you're planning to hide this all away? Let history continue to see them as frauds and failures?" She did not realise she cared so much how the world saw her parents. She had come through her own personal journey: from being embarrassed by and angry at them, to compassion for the panicked young scientists they must have been—trying to hide research from their corporate overlords for fear of the poor use it would be put to. Somewhere in there she had revised her vision of them as scientists. They weren't failures; they were geniuses so far ahead of their time that it was easy for the world to believe in the Hoax, even if synthesised humans did turn out to be possible several decades later. The thought that the world at large would never recognise that horrified her.

"We're not even sure this is real," said Tony. "With the Schwarz track record, it *might* just be a joke."

Carolyn fixed Tony with a glare, then turned to Dorsey in dawning realisation. "You haven't even told them, have you?" She waved her arm at her fellow scientists. "Where did they say this came from?"

"Hawaii," answered Jayla.

Dorsey stepped forward. "I think we've said enough here."

"No we haven't! This isn't your research to do with what you want! It's my parents'!" She was vaguely aware of a ripple of shock through her companions at the table, but her concentration was on Dorsey.

"They're not around to claim it. No one is. Vivcor ceased to exist a few days ago. It was found in a safe in government property in Hawaii."

Carolyn was suddenly distracted by the thought of that sad young man in the Vivcor basement in Bucharest; he was right, although it had been slightly more than a week before he no longer had a job. Anger filled her. The US government couldn't steal research done in Romania just because it had been hidden in Hawaii. It wasn't hidden only in Hawaii. "It was *born* in Bucharest," she said.

"Let's not argue about who owns the research," said Dorsey, his voice strained. His eyes flicked at the open door. "At the very least, it's not yours to say—"

"It could not be *more* mine!" she shouted. She spun to face the scientists. "There was no Hoax. I am Carolyn Schwarz. Baby C. All this,"—she waved her hand over the elly-books scattered on the table—"was encoded into my genome. All that was in a safe in Hawaii was a copy of my sequence." She turned back to Dorsey. "Ethan said they passed it on to me. They did so quite literally. I've got it with me all the time. Don't you dare tell me I have no say in what happens to their research!"

"Carolyn, you need to calm down." Dorsey stepped towards her, then away to push the door closed. He returned. "Just hear us out—"

"No!"

"Carol—" Dorsey cut off as the door slammed open again. "I'm sure she'll calm down," he said to the four agents who had entered. "These scientists have a delicate dispo—"

"Come us with us, please, Ms Schwarz," said the lead thug, taking hold of her arm. *Dr Schwarz*, she thought to herself furiously. She lunged and grabbed her elly-book with her other arm; she tucked it up to her chest. The men surrounded her, herding her out the door.

"Don't let them hide it!" she shouted to other scientists. "Don't let them leave the wrong history of the Schwarzes to stand!"

* * *

Carolyn sat on the bed in her room, feeling silly. Why had she done that? She had let emotions get the better of her. Now, instead of being part of things, she was locked in her room like a naughty child. She hugged her knees. She hoped the rest believed her. She hoped they believed her parents weren't frauds. She closed her eyes. She had to separate her emotional reaction from her current situation. What was she going to do next?

Her parents had gone to extraordinary measures—died, even—to keep their research from falling into the hands of a small group of people. This wasn't a corporation, but it was another small group. To ensure the safety of *all*—to keep some kind of ruling class of immortals from rising above the rest—the Schwarz Final Findings needed to be shared with everyone.

Chapter Forty-Seven

She scooted further under the covers, hiding the elly-book under a tent made from her knees. The soft cosy warmth of the bed sat in strange contrast to her tense excitement. At least she had set a pattern of reading late at night, tapping out notes on her elly-book, so if the room was under video surveillance, she would not look too unusual. She had had plenty of time during the day, with lunch and dinner delivered by silent agents, to plan what she was going to do.

She could not wait any longer. Perhaps she might be able to convince these other scientists to share the research, but she barely knew them. Any personal conversation had always been cut short. They might even be motivated to cooperate with the FBI because it meant they could get credit for her parents' breakthroughs. That was as bad as being a fraud, in her mind, but she did not know about in theirs.

She hoped that Susan was distressed enough by Carolyn's lack of contact to keep the tunnelling app going regularly. Or that she had recalculated midnight for Carolyn's new location. If her keepers had some way to detect generic internet communication, she might only have one shot at this. She needed to connect to Susan, give her the bare bones of the story, then get her parents' research out to as many outlets as she could.

She still vacillated over revealing that the secret was encoded in her genome; Ellen had half of it—would holding the secret put her in danger? But if everyone knew all of it, there would be no secret.

She breathed deep. She had to believe that spreading the information as widely as possible was the best way to keep them both safe. She would be limited by the browser's text interface: while it could deal with files, it was very slow. She had made zipped files of her genome and of her translation plus notes, a PDF file of the translation, and also planned to simply copy and paste onto websites. She had mapped out the path carefully, optimising locations she could find quickly and places that would get it to right people. If they came and stopped her, she wanted to have gotten as much out as possible.

She accessed the tunnel app and froze it at the cloud screen. She typed Susan's code, but the screen kept swirling. Susan wasn't there!

Then it stabilised, but back on the cloud page, a message in the middle saying, *Bridge Lb31 Requested. Accept? Deny?*

Confusion settled quickly into an uneasy queasiness. She'd seen that before—the mysterious other person, who had tried to contact her in Hawaii. Or perhaps someone else. She didn't remember the code. She didn't like the idea of opening a connection with a stranger.

But it was a connection. She needed the internet *now*.

If this was the antweb, like Susan had said, whoever that was should be anticorporate. Unless people like the FBI just sat around with elly-books requesting connections to random codes until one stuck. Yet would they have left her with her elly-book if they knew it was internet accessible? Her mind swirled, unable to process all the possibilities. Was she too paranoid? Not paranoid enough? She had been oblivious for so long, just trusting corporations, police, even Mike and Dorsey. She had no sense of what was reasonable.

She tried to apply logic. When she had chatted with Susan, neither could see what the other did on the browser half of the screen. She didn't have to tell the other person much. If she could keep a conversation going for long enough to get her message out, it would be worth it. Still, her fingers tingled in fear as she tapped 'Accept'.

Her screen swirled into the familiar split screen. At the top appeared, *Dr Gray????*

A shock of surprise flushed through her. They knew who she was.

She forced herself to ignore the message. She needed to make use of the internet connection. In the browser half, she called up bioRxiv, the biology pre-print server, where she had an account. The interface was different in the strange coloured text view, but she soon found the upload page where she titled her submission 'The Schwarz Final Findings as communicated by their actually-synthesised daughter, Baby C'. It would hold her account name of Carolyn Gray, but that brief news report she had seen in Uncle Keith's conservatory had connected her identities. Given she was supposed to be dead, she hoped someone would take the time to read what she had posted from 'beyond the grave'.

If someone in those hidden halls of corporate or government power didn't get it deleted first. Which was why the last item in a message she had prepared for Susan, explaining the last few days, was to ask for addresses of anticorporate message boards. But she didn't have Susan on the other side of the tunnel. As the upload progressed, she turned her attention to the top screen, which now also said, *It's you, isn't it? I KNEW you weren't dead. Dr Gray? Please answer me.*

Who are you? she typed. If she didn't engage, the other person might end the connection.

I'm your student, Kelsey Brown. You've got my elly-book.

Kelsey. The brown-haired woman who had approached her that last lecture, hugging an old tablet. But how could she be sure? She was suspicious of everything, now. *Why do you think I'm Dr Gray?*

If text could seem hesitant, the slow reveal of the next words gave that impression. *My elly-book was in your flat. Or near. At least. Then in the antco centre of Bucharest. Something's going on. I don't like it.* The text's hesitancy vanished. *They got to Juliet. She's not here anymore, but I don't think she's spilled. I know there was no prank because we were together that night. I kept silent because no one but me and Juliet knew. But my elly-book was gone and it was in your flat and it could have been taken and I was so silly and left material on it I shouldn't have but then nothing happened and then it was in Bucharest and in the antco and I knew there was no prank and then I guessed there was no death and that you were alive but running. From whatever Juliet ran from. She's gone, and*

I don't think they have her because I'm here and the rest of us are here and they move faster than that.

She stared at the thick block of text, barely able to parse it. The bioRxiv loading finished. She filed her confusion away and returned to the browser. She navigated to and tapped on the 'Submit' button. One confirmation page later, where she declined to 'View her submission', she moved on to her next science website. She had only two more planned before she was hoping to get to anticorporate sites—which she didn't have because she didn't have Susan.

She read the previous text message again. She took a deep breath. She would risk it. *I have material I need to get out. Do you know some anticorporate message boards with wide reach?*

Yes. What do you want to get out? Now the other side of the connection seemed suspicious.

You read the article? That I'm Baby-C? She was implicitly agreeing with Kelsey's identification of her, but she had already decided to take the risk.

Yes.

The Hoax was a Hoax. I'm synthesised. The Schwarz Final Findings were real. I have them, and they have to get out before we repeat all this over again. Would Kelsey even know about the Schwarz Final Findings? Probably not. It was discredited, old news, well before the young student was born.

The screen showed only, *OMG!!!* Carolyn continued her uploading, moving quickly through the other top-priority science sites. She waited, tense, wondering when the door might burst open and someone demand what she was doing. But that did not happen before three web addresses appeared in quick succession.

Thanks, she took the time to type, before entering the first of them. It allowed her to register for an account simultaneously with posting a message, so she titled her message the same as her pre-print submission and pasted in the translation. The other message boards took more time to register, but soon two more messages were posted. The third allowed her to upload attachments as well.

She hesitated, but she had already thought this through. She made a new message and attached her zipped genome. She watched the slow upload begin, then turned her eyes back to the chat window.

Wow, it said. *That's amazing. I even understand some of it from class! :)* Kelsey must be reading her postings.

Carolyn moved on to her lower-priority science sites. She kept going, getting further than she had imagined or hoped. Excitement started to build. Perhaps the connection was passing unnoticed. She could have all night. Next on her list was her email. She had prioritised it lower because it might take some time to see if she could still log in, but if she could send personal messages to people they would be harder to track down and delete. She started navigating the university site.

She flicked her eyes up to the chat. *Dr Gray, are you okay?* it said.

Was she okay? She was not injured. But she had not thought beyond uploading as much as possible before the FBI stopped her. Yet they hadn't and might not realise until morning, when she hoped the message would have gone world-wide. They would not be happy with her, she was sure. Would they harm her? She didn't *think* so. But she had not thought they would keep her captive like this, either. Dorsey seemed on her side, somewhat, but he clearly wasn't in charge.

She looked to the window and the sleepy suburban neighbourhood beyond. Given her success this night, perhaps surveillance was less than she had thought. Could she just walk out? Yet where would she go? She didn't even know where she was!

But she might be able to find out. *I'm okay, but I could be in trouble soon. Can you help me?*

How?

I need to know where I am. Can you site your elly-book?

Location services have been off since I found you in Bucharest.

She was trusting this person. There no reason to believe it wasn't Kelsey. She took a deep breath, then typed, *How do I turn it back on?*

By the time Kelsey had walked her through turning on location services and sent back an address, Carolyn had emailed everyone in her address book both her parents' findings and her genome. There were some more science sites she had thought of, but with her emails out there, she may have done enough. If she could get *herself* out, she could do more later.

What she really needed was a contact in the anticorporate world in … she checked her address … Virginia. She vaguely knew Virginia was on the east coast of the US, but that didn't help much. The town, Reston, meant little to her. *Do you know where that is?* she typed. *Do you know any anticorporate contacts here?*

You're just outside of Washington, DC., Kelsey wrote back. *Near Dulles airport.* That made sense. They hadn't driven very far, either to that first safehouse or to here. *But I don't know anyone in that area. I can find out, but it might take a few days.*

She didn't have a few days. She had a few hours. The only people she knew here were Mike and Dorsey. Dorsey was at worst part of her problem, at best useless. Mike … Mike might be helpful. That he had passed on his phone suggested he guessed she might want to get in touch. But he obviously hadn't guessed they'd take it away from her!

Susan had found nothing about Mike when she looked for him, with her police-savvy and from an anticorporate centre. Kelsey, whatever her connections with the anticorporate world, couldn't do more any faster than she could just find her own contacts. Which was days. Carolyn needed something now. If only she could get in touch with Mike! But he was a ghost.

A ghost with a sister—an anticorporate sister—she suddenly remembered. *I need to get out tonight. There's someone who could help—his name is Mike Hafal, but he'll be hard to find. He has a sister in the anticorporate community. Her name is—*she screwed up her face, trying to remember the name she had heard from Leia in Hawaii, just a few days ago, was it Jill? No—*Jenn. Would you be able to find a Jenn Hafal?*

I can check public phone books, but finding someone through the antco would take longer. There was a brief pause. *No Mike or Michael Hafal, or Jenn, Jenny or Jennifer Hafal.*

I'm not sure the Jenn is in this area.

There's no such Jenn in the whole US, at least in the public databases.

Her heart dropped. But did she have the surname correct? Women still sometimes changed their names on marriage, especially in the US. And Mike could have lied about his surname, as he did originally to Ethan. She tried to remember what Ethan had thought his name was. *Try Jenn Talon.*

There's over 800 Jennifer Talon, Tallon or Talen in the US, 22 of them in Virginia.

That was the opposite problem. She couldn't contact them all! Plus, the chances of Mike using his real name as his pretend name, in some kind of double-bluff, were slim. All she had was a first name. Twenty-two Jenn Talon's in Virginia alone; there would be vast numbers of Jennifer's with any surname. Did she know anything else that could help her? If only she had Mike's phone!

Mike's phone: Jenny-heart.

Her own heart raced. Maybe Mike hadn't mentioned a partner because he didn't have one. What if Jenny-heart was his sister? *Try Jenny Beane.*

There are two Jennifer Beane's in Virginia. One is in Herndon, that's right next to you. The other is in Arlington, not far away either. The one in Herndon is 92 years old. The one in Arlington is 30.

Grandmother and granddaughter? *Can you get in touch with the one in Arlington for me?*

Chapter Forty-Eight

Carolyn opened her window in the dim pre-dawn light. It had been a tense few hours, pretending to sleep while waiting on Kelsey's sparse communications and fearing a pounding at her door. She did not pack ahead of time, not wanting to tip off anyone who might be watching her room overnight. Luckily her belongings were few, and she had been able to shove them quickly into her knapsack.

By the same logic, she should not stand long, fully dressed in her coat, at the open window. She bent over and stuck a leg through the window. She swung her other leg out, hung on her stomach over the sill for a moment, then dropped the fraction of a metre to the ground. Dry rhododendron leaves crunched her feet. She pulled the window closed. She walked quickly along the side of the building, until nearly to the next window, then struck out perpendicularly.

There were no sounds of pursuit. She kept going, walking across manicured grass to the winding suburban street, lit only by the widely separated street lights. She reached the street and turned left, following the directions passed along from Jenny Beane via Kelsey.

Despite everything, her brain could not help revisiting an already worn path: Jenny-heart was Mike's sister. Excitement surged. Her interactions with Mike spun past: if she had not assumed a partner, how might those events be interpreted differently? Could he, possibly, impossibly, be interested in her?

She reached a larger main road, marked by one of the iconic eight-sided stop-signs she had seen in so many American films. She turned right and walked uphill along the main, windy road. Her back tingled. She looked behind her; the community centre-like building was nearly out of sight. She pushed aside thoughts of Mike, instead concentrating on her instructions.

After three roads on the right, cross to the parking lot through the woods on the left side of the road. A sculpture like 'a pile of sugar cubes' marks the entrance to a shopping plaza. Down two sets of stairs she would find a lake with several rowboats tied along the dock. The blue one has a tracker in it. Get in it and out into the lake; Jenny would find her there.

Was that right? Three streets? Or was it four? She was sure about the blue.

She almost thought she heard a sound. She looked back; bright lights shone where none had been before. A yellow strobe emanated from where her recently vacated building would be, just out of sight. Fear clenched her heart. They had noticed her escape!

She could not just *stand* here. She ran, only belatedly remembering she was meant to count streets. Was that three? She dashed across the street into the woods, hoping she had not gone too far. A stitch stung in her side; she slowed to a walk, gasping, loud in the dim. She looked behind her. She could see no lights, not even the street, only silent silhouettes of trees. She walked forward. Dry leaves made soft crackles. Morning birds began to sing. Just before she was sure she had entered an unchartered wilderness somehow inserted near the US capitol, she broke through into a parking lot. She had counted correctly! Scattered cars parked, mostly near the left edge. She wound around the outside of it, past the cars, towards a series of brick buildings. She hurried down a walkway broken in the centre by a shadowed, blocky sculpture, hoping that it was the 'pile of sugar cubes'.

Tyres crunched on asphalt in the parking lot. Fear spiked again. No one should be out at this time. She ran. Down a set of wide stairs, then another, and she found herself on a small jetty at a lakeside that would have been a surprise had she not been expecting it.

"Do you have a live feed?" said a male voice, from somewhere behind her.

"No, they're not on the net!" an annoyed female voice replied, some distance from her companion.

"I thought I saw someone head to the lake," said the man.

"That's a dead end. If she's there, she'll keep. Check the shops first."

Carolyn knelt and examined the row boats. She could not tell colour in the dim morning light. Too dark; too dark; that one was white. Let this one be blue, she thought. She folded back the tarpaulin covering the boat and stepped in. She turned her attention to the rope holding it to the dock. It was wrapped in a complicated pattern around two small bollard-like items.

She leaned forward and started unwinding it, her breath seeming so loud she was sure those two people would come running. But she finally released it, picked up an oar, and gave a mighty shove against the dock.

The boat leisurely crept back a few tens of centimetres.

She couldn't row; that would be too loud. She gave another shove with the oar. She heard—or imagined she heard—the voices again. The woman was right; this was a dead end. She would have to hide.

She lay down in the boat and pulled the tarpaulin back over. The confined space smelled strongly of mildew. The bottom of the boat was wet; cold water seeped into her hair. She bit her lip, fighting tears. She had been so close to freedom!

Footsteps on stone; then the hollow sound of two people walking on the dock. "Are you sure she came in here?" said the woman.

"No!" The man's tone was one of exasperation. "But where else could she have gone?"

"She could be hiding in the woods! Up a tree! I don't know what a scientist would do." The woman's tone matched the man's. "Why did she go out the window in the first place?"

"Who knows why a scientist does anything?" said the man.

There was a humongous rattling sound, and Carolyn's boat swayed. Someone must have kicked a boat.

"Let's head back and check what vid we do have. Not sure why they care, anyway," said the woman. "I heard she wasn't helping anymore."

"Apparently she didn't sign the same thing the rest of them did," said the man.

"How'd they let that happen?"

"Dunno. Not our issue anyway."

Their footsteps left. Carolyn closed her eyes, squeezing out the tears she had resisted before. She stayed frozen, fearing they would come back. By the time she pushed the tarpaulin back, her back was soaked through.

She sat up and looked around in confusion. The jetty and shops were gone. She saw only water and trees, with red sunlight streaming through the branches. Dawn. She must have lay there long enough for the boat to have drifted into the lake on its own.

She squinted at the paint on the bow. Her heart dropped. It was green. Or was it turquoise?

A distant roar resolved to a wake nearly obscuring a motorboat coming from around a forested prominence. She stared at the approaching vehicle, frozen in mind and body. She was cold, wet, and tired. She barely cared what happened at this point.

The boat slowed. Its wake sunk, revealing the silhouette of a man and woman on board. "Do you call that blue?" Mike's voice asked.

Relief flooded her. "I couldn't tell colour in the dark."

"Oh. Sorry," said the woman. She was the same height as Mike, and the family resemblance was obvious.

Mike grinned. "It's all right now." He leaned from his boat as it settled alongside hers. He held out a hand. "You definitely made an impact out there."

"They were going to hide it away again!" she said, ready to plunge into an explanation. She took his hand and stepped onto their boat. She looked at the rowboat. "Should we—"

"Don't worry about it. Someone will fetch it." Mike's grin was infectious. Carolyn felt her heart lift. "I don't think anyone will be able to hide away the Schwarz Final Findings now."

"Did I get the message out?"

"Oh, you got it out all right," said Jenn.

"Europe woke up to your emails in their inboxes hours ago, and the shock is slowly making it across the ocean." He looked around. "But this isn't quite the place. Let's get you dry and inside."

Jenn took the controls of the boat as Mike helped Carolyn from her wet jacket and wrapped his own dry one around her. He kept his arm around her. She leaned into him.

"You're going to have to tell me how you managed all that, by the way," he said.

She closed her eyes, nearly about to drift off right there in the middle of the lake. She opened her eyes and looked at Mike, his face not half a metre away. "I can't get a new identity now, can I?" She had splashed her not-death and her not-a-hoax identity across the world.

"You could," he said. "Do you want one?"

Did she? What would she do, pick up doing science for corporations again? She couldn't do that. She had to make a change in the world. She already had. The world now had her parents' longevity treatment. It would take some time to confirm, but she was sure it was solid. They had not been frauds. She needed to finish the job, to make sure the longevity treatment did not become the property of some small group. She needed to be part of making sure the stranglehold of corporations was broken. "No," she said. "Or rather, yes, but just a little. I think I'll be Carolyn Schwarz again from now on."

Epilogue

Carolyn looked over the faces arrayed before her: eager graduates, proud parents. She shifted in her heavy American-style academic robes, then stilled, remembering she was being filmed. Thirty years of being a figurehead of the Free Science Movement—risen to the very top of the academic world—and she still found public events like this nerve-wracking. But she was considerably better at not showing it.

Her gaze darted to Ethan, sitting politely unruffled as the young-looking professor at the podium continued her commencement address. His robes echoed her own, but with black highlights on green instead of gold, representing his position as President of the College of William and Mary. Ethan appeared aged, but not as much as the passing years accounted for: he was among the first recipients of the experimental treatment that Carolyn's parents had encoded—and it was now clear, also expressed—in her genome.

Carolyn had heard the adage that students looked younger as your career progressed; but it was now not so much the students but the parents that looked younger. And the appearance was not psychological, but physical.

These graduating students were launching as adults into a world so different than the one she had known at that age: the worries of the aging population would soon be replaced by the worries of the non-aging population. What careers would these young people aspire to when people their grandparent's age—Carolyn's age—continued in the prime of life? Perhaps not so much this generation, with spotty uptake of longevity treatment in a suspicious populace, but those to follow: society would need to change so much more than it already had.

Applause grounded her back in the present. Ethan re-took the podium as the applause died down. "Thank you Professor Brown for those stirring reflections on our changing world." He paused to shuffle papers in front of him: entirely for effect, Carolyn suspected. "It is my great honour to move onto the next element of my final graduation ceremony as your President. Before the conferral of degrees—both esteemed recipients of Honorary degrees I've had the great pleasure of meeting these last few days, and hard-working recipients of our Bachelors, Masters, and Doctorate degrees I've had the even greater pleasure of knowing as members of our student body,"—a sweeping gesture encompassed the students in front of him—"I am honoured to swear in the second Chancellor of William and Mary since my tenure as your President."

He turned his head briefly towards Carolyn, then bestowed an avuncular smile on the hall at large. "As you will know, the College of William and Mary in Virginia under its Royal Charter looks to a Chancellor for the intellectual and spiritual leadership of the College. The Board of Visitors has seen fit to vote as our next Chancellor, an old friend of mine, Professor Dame Carolyn Schwarz." He paused for the swell of applause. He flashed Carolyn a quick grin, then returned to his speech. "Professor Dame Schwarz is the current Principal of the Free University of Scotland and is known throughout the world for her tireless efforts to return, then maintain, Free Science as a driving force in our society. Of course, she is also known for her ground-breaking work in uncovering, confirming, and expanding upon the research of her parents, Drs Schwarz and Schwarz, showing the that mythical Schwarz Final Findings were first, far from mythical, and second, far from final, at least when counting the next generation of Schwarz …".

Carolyn forced herself to continue to pay attention to Ethan's introduction, despite the public singing of her praises being one of her least favourite aspects of her career. When he finally stopped speaking, she approached him and bowed her head to receive the chain and medal of her new office. Ethan clasped both of her hands in his warmly, then guided her towards his previous location behind the podium.

She stood straight, the unfamiliar weight of the chains and medal settling across her shoulders like a pair of arms, embracing her. She smiled at the comforting image. Standing on this foreign soil of the country where her parents had met their untimely deaths—whose cold case had been finally solved and the corporate culprits brought to justice—felt a strange kind of full circle. "Thank you, Professor Boltzer—Ethan—for that amazing introduction and thank you, to the Board, for your confidence in me. I look forward to working with Ethan in this final year of his Presidency. And, as Ethan is outgoing and thus less vulnerable to the embarrassment I may be about to cause ..." She paused to let a light, slightly uncomfortable laugh pass through the crowd. Ethan smiled, and the laugh continued more relaxed. "I hope he does not mind if before I move onto my prepared speech, I share a little story about how the two of us met ...".

Part II

The Science Behind the Fiction

2

The Biology Behind Carolyn's Code

Introduction: The Complexity of Genomes

Before plunging into the details of how Carolyn's parents made her and what, precisely, her code was, we'll start with some basics on DNA and genomes.[1] Most people have heard of the 'double helix': the three-dimensional structure of DNA. DNA stands for *deoxyribonucleic acid*. A DNA molecule is built up of smaller subunits known as *nucleotides*: each nucleotide consists of a sugar, a phosphate group, and a nitrogenous base (Fig. 2.1a). The sugar is deoxyribose—in DNA's name—and the phosphate group is negatively charged making the molecule an acid. When many nucleotides bind together, these two elements make up what is known as a sugar-phosphate backbone (Fig. 2.1b); two backbones wind around each other generating the double helix (Fig. 2.1c). Between these two backbones are ladder-like elements formed of the bases. These bases are the variable part of DNA: what is typically termed 'the genetic code'. There are four types of bases in DNA: adenine (abbreviated A), guanine (G), thymine (T), and cytosine (C). The order of these bases in consecutive nucleotides of the DNA molecule are its *sequence*. The size of a DNA molecule is measured in *base pairs*, abbreviated *bp*, referring to the number of paired bases across the two sides of the helix.

The key to DNA's ability to transmit the genetic code—the DNA sequence—from one generation to the next is that each nucleotide in the

---

[1] Much of the material covered in this essay can be found in any good introductory biology textbook, for example Campbell et al (2017) *Biology: A Global Approach* 10th Edition, Pearson Education Ltd: Essex, UK; where particular other sources are of interest, pointers will be provided.

© Springer Nature Switzerland AG 2019
V. A. Smith, *A Code for Carolyn*, Science and Fiction,
https://doi.org/10.1007/978-3-030-04553-1_2

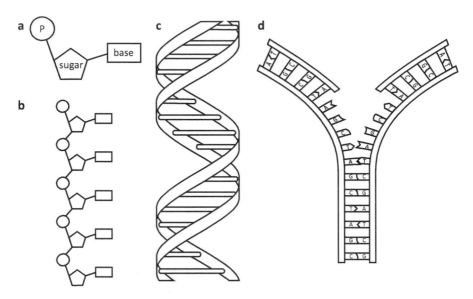

Fig. 2.1 DNA structure and complementary base pairing. (**a**) Schematic drawing of a DNA nucleotide, showing the sugar, phosphate group (P), and nitrogenous base. (**b**) Successive nucleotides join by binding the sugar of one to the phosphate group of the next, resulting in a sugar-phosphate backbone with bases hanging off on one side. (**c**) Schematic of the three-dimensional structure of DNA, showing two backbones winding around each other in a double helix, with the bases joined across the centre like rungs on a ladder. (**d**) Complementary base pairing enables exact copies to be made of a double-stranded DNA molecule: the two strands separate and new nucleotides match to their complementary bases, recreating the missing other side

double helix binds base-to-base to another nucleotide on the other side: the bases adenine and thymine always bind, as do guanine and cytosine. This is known as *complementary base pairing*. So if one side of the helix—also known as a *strand*—contains A, it immediately follows that the other strand contains T in the same place. In this way a DNA molecule can always generate an exact copy of itself: the two halves of the double helix need only separate and build back their missing halves, recreating the same sequence in each (Fig. 2.1d). This is what happens every single time a cell divides: each cell in our bodies has a full copy of our entire genetic code. We'll explore later in this section what happens when transmitting DNA from parent to child, and why everyone thought Carolyn was not artificially synthesised.

From Gene to Protein

But for now, let's look more closely at this genetic code. What does it consist of? Popular knowledge tells us that 'genes' are what is passed through generations. This is true: the genetic code contains genes, and more. A gene is a section of DNA that *codes for* a protein. This was the basis of Carolyn's and Mike's first conversation: the one thing (other than gardening!) that Mike knew about biology was that genes were based on a *degenerate code*. This means a code where more than one set of symbols map onto the same outcome: in the case of DNA, this is a set of three bases, or a *codon*, which corresponds to an *amino acid* in the protein. A gene consists of many codons, which tells the cell in which order and how many amino acids to use to make a protein (Fig. 2.2). Like nucleotides make up DNA, amino acids make up proteins. Proteins are longs strings of amino acids that fold back over themselves into all sorts of useful and interesting shapes. Proteins are what "do" most biology: they catalyse the chemical reactions that allow us to live (*enzymes*), they form major parts of our physical structure (*cytoskeletons*: what gives a cell its shape), they act as messengers between parts of our body (*hormones*), and so on. Thus, our genes, coding for our proteins, are a significant part of the genetic code.

first base	second base: T	C	A	G	third base
T	TTT Phe (F)	TCT Ser (S)	TAT Tyr (Y)	TGT Cys (C)	T
T	TTC Phe (F)	TCC Ser (S)	TAC Tyr (Y)	TGC Cys (C)	C
T	TTA Leu (L)	TCA Ser (S)	TAA STOP	TGA STOP	A
T	TTG Leu (L)	TCG Ser (S)	TAG STOP	TGG Trp (W)	G
C	CTT Leu (L)	CCT Pro (P)	CAT His (H)	CGT Arg (R)	T
C	CTC Leu (L)	CCC Pro (P)	CAC His (H)	CGC Arg (R)	C
C	CTA Leu (L)	CCA Pro (P)	CAA Gln (Q)	CGA Arg (R)	A
C	CTG Leu (L)	CCG Pro (P)	CAG Gln (Q)	CGG Arg (R)	G
A	ATT Ile (I)	ACT Thr (T)	AAT Asn (N)	AGT Ser (S)	T
A	ATC Ile (I)	ACC Thr (T)	AAC Asn (N)	AGC Ser (S)	C
A	ATA Ile (I)	ACA Thr (T)	AAA Lys (K)	AGA Arg (R)	A
A	ATG Met (M)	ACG Thr (T)	AAG Lys (K)	AGG Arg (R)	G
G	GTT Val (V)	GCT Ala (A)	GAT Asp (D)	GGT Gly (G)	T
G	GTC Val (V)	GCC Ala (A)	GAC Asp (D)	GGC Gly (G)	C
G	GTA Val (V)	GCA Ala (A)	GAA Glu (E)	GGA Gly (G)	A
G	GTG Val (V)	GCG Ala (A)	GAG Glu (E)	GGG Gly (G)	G

Fig. 2.2 DNA codons and their amino acids. Each set of three DNA bases, or codon, refers to a specific amino acid, or a stop signal, for making a protein. The codons are degenerate: more than one codon can specify the same amino acid. The codons are shown alongside their amino acid or the signal to stop. The twenty amino acids are written as their three-letter abbreviations: *Ala* alanine, *Arg* arginine, *Asn* asparagine, *Asp* aspartic acid, *Cys* cysteine, *Glu* glutamic acid, *Gln* glutamine, *Gly* glycine, *His* histidine, *Ile* isoleucine, *Leu* leucine, *Lys* lysine, *Met* methionine, *Phe* phenylalanine, *Pro* proline, *Ser* serine, *Thr* threonine, *Trp* tryptophan, *Tyr* tyrosine, *Val* valine. The single-letter abbreviation for each is also shown; the single-letter abbreviation is what was written on Carolyn's parents' ID cards

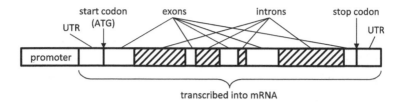

Fig. 2.3 The anatomy of a gene. A schematic layout of a gene's sequence as found on DNA is shown (not to scale). The first element is its promoter, which regulates how much the gene is transcribed. The remainder is transcribed into mRNA; sometimes only the transcribed part is preferred to as the gene (and the promoter is considered to be placed 'before' the gene on the DNA). The elements of the resulting mRNA are shown here: untranslated regions (UTRs) before the start codon and after the stop codon, and exons—translated into proteins—and introns—cut out during mRNA processing

A gene is a little more complicated than just a bunch of codons, however. Figure 2.3 shows the 'anatomy' of a gene. A gene starts with a *promoter*: this is a sequence in the DNA that is recognised by proteins that control whether a gene is "on" or "off", and if "on", how much. Regulatory proteins will *bind*, or stick to, the promoter region in the DNA. They then either recruit (if turning "on") or inhibit (if turning "off") the binding of what is known collectively as the *transcriptional machinery*. The transcriptional machinery is a big pile of proteins that ultimately result in one kind of protein, known as *RNA polymerase*, running down the remaining length of the gene and making a 'copy' of the gene in a molecule that is very similar to DNA: *RNA*, or *ribonucleic acid*. RNA has three main differences from DNA. First, its sugar-phosphate backbone consists of the sugar *ribose* instead of *deoxyribose*[2]; second, it exists in a single strand instead of the double-stranded double helix of DNA; and third, where DNA uses the base thymine, RNA uses a different base, known as uracil (abbreviated U). The process of making RNA from a section of DNA is known as *transcription*. The result of transcription is a piece of RNA known as *messenger RNA*, or *mRNA* for short.

The mRNA that is the exact copy of the remainder of the gene beyond the promoter is often called *precursor mRNA*, or *pre-mRNA*, because there are a number of things that happen to it before its code is ready to undergo *translation*, the process of making a protein. Pre-mRNA starts with a short bit of sequence known as an *untranslated region (UTR)*: this sequence does not give

---

[2] The only difference between these two is that ribose has one more oxygen than deoxyribose: the 'deoxy' part of the name just says it is ribose minus an oxygen!

instructions for amino acids in a protein, and instead binds proteins which control the process of translation. The start of what is known as the *coding sequence* is always signalled by the codon for methionine (this is what clued Carolyn and Ethan into realising they were looking at translated coding sequences on her parents' ID cards). What follows is a pattern of sequences that will be translated into protein, known as *exons* (for expressed regions), and sequences that are not translated into protein, known as *introns* (for intragenic regions[3]). The final exon ends with what is known as a *stop codon*: this is one of three codons that do not code for any amino acid (Fig. 2.2). After the last exon is another untranslated region; this part contains sequences which influence a variety of features, for example, where the mRNA goes in the cell, how long the mRNA lasts, and how fast it gets translated into protein.

Before a pre-mRNA gets translated into protein, it undergoes *processing*, to turn it into a *mature mRNA*. This involves removing the introns from between the exons, so that all that remains is exactly the code for a making a protein (the *coding sequence*: this, translated into the letters corresponding to amino acids in the resulting proteins, is what was written on Carolyn's parents' ID cards). It also involves adding some sequence before the first untranslated region, known as a *cap*, and some after the final untranslated region, known as a *tail*. Both of these elements help stabilise the RNA molecule and provide further signalling information for the rest of the cell about what to do with this specific mRNA. The mature mRNA then binds both a *transfer RNA*, or *tRNA*, for methionine (remember, always the first codon) and a *ribosome*, a piece of cellular machinery made up of both protein and RNA. As Carolyn explained to Mike, each codon relates to a tRNA. A tRNA has at one end the complementary sequence for its codon, and at the other end is bound to the amino acid for which the codon codes. The remainder of the protein is built by the ribosome connecting the amino acid of each subsequent tRNA, complementary to each following codon, to the amino acid that has come before. When the ribosome reaches a stop codon, which has no complementary tRNA, the process stops. This is not quite the end: after all the amino acids have been strung together, there is often what is known as *post-translational modifications*. These are changes made to the protein after it is translated, one of which may be chopping off that very first methionine—knowing about this is what made Carolyn, in her inherent drive for accuracy, specify that the

---

[3] The term 'intron' originally came from 'intragenic region', meaning region inside of a gene; however, it is commonly now also interpreted as 'intervening region'.

letters on the ID cards were translated coding sequences only (and not necessarily the final protein).

"It's Not Junk"

The genes themselves only make up about a quarter of our genome—and that quarter is heavily biased towards introns, which don't actually code for protein. Overall, exons make up ~1.5% of our DNA, promoters and other regulatory sequences ~5%, and introns ~20%. So, what is the remaining three-quarters of our DNA doing (and why do our genes consist mostly of stuff that doesn't actually code for protein)?

For a long time, people called the non-gene portions of our DNA 'junk DNA'. But that view is slowly changing. A project called ENCODE (**Enc**yclopedia **of D**NA **E**lements) found that 80% of human DNA has 'biochemical function'—as a regulatory sequence or as the ~75% of the genome that they found transcribed into RNA in at least one cell type. The project's leader even suggested that had more different types of human cells been examined, they may have found function for all of our sequence![4] While such predictions (and their interpretation) are still a contentious area, the fact remains that a lot more of our genome is doing *something* than originally thought.

One chunk of this 'other' DNA consists of DNA sequence that appears to be detritus of evolution, for example fragmentary pieces of genes and *pseudogenes*: coding sequences that have accumulated mutations and are either no longer transcribed (as their promoters are disrupted) or produce non-functional proteins because their mutations are too severe. Some of these do have function in their RNA form, for example by binding to and interfering with translation of the actual mRNA of a related functional gene.[5] One of the most famous pseudogenes in humans is for an enzyme that can produce Vitamin C: humans have only a non-functioning pseudogene for this enzyme, which is why we need to ingest Vitamin C in our diet.

Most of the remaining DNA is *repetitive sequence*, the same DNA sequence repeated hundreds or thousands of times. Some of the repeated sequences are short, from 2 to 100s of base pairs in length. *Short tandem repeats (STRs)*,

[4] E Young. 2012 Not Exactly Rocket Science: 'ENCODE: the rough guide to the human genome', *Discover Blogs* (September 5, 2012 1:00 pm), Available at: <http://blogs.discovermagazine.com/notrocketscience/2012/09/05/encode-the-rough-guide-to-the-human-genome/> [Accessed 15 June 2018].
[5] MJ Milligan & L Lipovich. 2015. Pseudogene-derived lncRNAs: emerging regulators of gene expression. Frontiers in Genetics 5:476, doi: https://doi.org/10.3389/fgene.2014.00476.

repeats of the same 2–6 letter sequence, are commonly used in forensic science because individuals will differ in the number of each type of repeat they have; law enforcement agencies use panels of around a dozen STRs that can specifically identify an individual. Currently, it still takes quite some time for the laboratory work necessary to count these sequences, but STRs are probably what Ethan was referring to when he talked about the gene scan devices that had supplanted mitocyls. Mitocyls are also a science-fictional invention, which as described by Ethan are capable of identifying either a few SNPs—*single nucleotide polymorphisms*, a single different base in DNA—in a mitochondrial sequence or identifying a specific individual (one presumes by a larger panel of SNPs); it is reasonable that technology for identifying SNPs would be available on quick timescales before the more involved STR-length determination.

By far the most numerous (and potentially most interesting) kind of repetitive DNA is related to what are known as *transposable elements* or *transposons*. This type of repetitive DNA makes up nearly half the human genome. Transposons are found throughout all domains of life, from bacteria to humans. They are pieces of DNA that move about in the genome, sometimes leaving copies of themselves behind. The most basic of bacterial transposons consist of nothing but flanking 'signal' sequences surrounding an enzyme; the enzyme recognises the flanking signal, cuts it and the intervening DNA out of the genome, and inserts it somewhere else. Transposons in multicellular organisms can be considerably more complex, including features such as copying into RNA as part of the transposition process and carrying genes or exons around with them. Transposons that only copy themselves, instead of moving their entire sequence, can end up leaving many copies in the genome, thousands or even millions. One type of transposon alone, known as *Alu* elements, exists in ~1 million copies and makes up 10% of our genome!

Transposons are often portrayed as 'parasitic': by making copies of themselves in our DNA, they are transmitted down generations as we transmit all of our DNA. Some transposons have even lost the enzymes for copying themselves, retaining only the signal sequences, making them parasitic on other transposons for their transposition. Transposons and viruses are thought to be evolutionarily related, although 'which came first' is still debated (and may not be absolutely one or the other).[6]

---

[6] EV Koonin, M Krupovic & N Yutin. 2015. Evolution of double-stranded DNA viruses of eukaryotes: from bacteriophages to transposons to giant viruses. Annals of the New York Academy of Sciences 1341:10–24, doi: https://doi.org/10.1111/nyas.12728.

Whatever their origin, transposons have also been co-opted by cells for various functions. In particular, our DNA is organised into long, linear sections known as *chromosomes*. Many elements of chromosome structure utilize repetitive sequence with their origins in transposons. For example, centromeres (the centre of chromosomes) consist in large part of transposon-related repetitive sequence. We'll see more of centromeres in the next section.

Before moving onto chromosomes (and exploring why everyone thought Carolyn wasn't synthesised), we'll go back and look at one last piece of the make-up of our genome: the introns. Unfortunately, it is not quite possible to definitively answer the question posed at the start of this section about introns; however, we can make some educated guesses as to what all this non-coding sequence inside our genes is doing, much based on the information presented above. First, one function of introns is to delineate exons: however, not every mature mRNA includes all the exons in the pre-mRNA. Genes sometimes 'mix-and-match' different bits of themselves to produce proteins with slightly different functions. Thus, introns enable the cell to make multiple proteins from the same gene. Second, much sequence in introns forms regulatory function (some of that ~80% of the genome with 'biochemical function' determined by ENCODE). And finally, introns may contain transposons. In fact, sometimes transposons can span across introns to include exons inside of them! When these transposons move or copy themselves, they take the exon with them, and potentially provide new function to another gene.

We continue to find more and more about the make-up of our genome, with surprising revelations of unexpected function. Like much of biology, the process is fractal-like: the more closely one examines something, the more detail appears. It is likely that this process is still continuing even by Carolyn's time: genetics is complex enough that they may be making 'designer babies' by modifying genomic sequences whose influence they do understand, while faithfully copying the remainder (and hoping that their modifications don't cause too many problems with yet-to-be-discovered interactions).

Chromosomes and the Human Hoax

It was Carolyn's karyotype, or chromosomal makeup, that led everyone to call her the 'Human Hoax'. A chromosome is a single DNA molecule. Chromosomes also contain proteins, about which the strands of DNA wind and which hold the chromosome into a complex three-dimensional shape. Chromosomes differ between *eukaryotes* and *prokaryotes*. We care here most about eukaryotic chromosomes (humans are eukaryotes), but we will say a few words about prokaryotes because that will relate to later topics. Eukaryotes include fungi, plants, and animals. They are characterised by cells that include complex internal structure with different functions occurring in what are known as *organelles*: sections of the cell subdivided off with membranes. Prokaryotes are distinguished from eukaryotes by lacking these organelles; all are single-celled organisms (bacteria are prokaryotes). Each prokaryotic chromosome is a single circular molecule, whereas eukaryotic chromosomes are linear: a molecule of DNA with two ends.

We'll take a look at the linear nature of eukaryotic DNA at the very end of this essay; for now, we'll concentrate on how a multicellular organism like a human generates the next generation. This occurs in *germ cells*, the cells that give rise to *gametes*—in animals, eggs and sperm—which are combined to make a new organism. The key to understanding why Carolyn was termed the Human Hoax lies in understanding what happens in these germ cells: how gametes are made. Before that, however, we need to cover the simpler process of how a cell divides, known as *mitosis*. There are a large number of biological processes involved in cell division, but we are going to focus on what happens to the DNA in its chromosomes. When a cell divides, its DNA is first copied. The two copies of an individual chromosome are held together by proteins binding to their centromeres. At a stage of mitosis known as *metaphase*, all the duplicated chromosomes line up in the centre of the cell. Then, in *anaphase*, the two copies of the chromosome go to opposite sides of the cell. Finally, the cell pinches apart into two pieces, each with one copy of each chromosome (Fig. 2.4a).

The DNA of all animals, except for some male insects,[7] is organised into pairs of chromosomes: we all have double copies of each piece of DNA. Parents give one copy of each of their chromosomes to their offspring, making them a genetic blend of their parents. This happens by combining a gamete from the mother (egg) with one from the father (sperm): unlike every other cell in our body, gametes have only one copy of each chromosome. When they join, the resulting *zygote* (fertilised egg) again contains double copies of everything.

The process of generating these gametes from germ cells is known as *meiosis*. Meiosis begins similarly to mitosis, by copying all the DNA in the cell, then goes through similar phases to mitosis, except twice—with some very

---

[7] Detailed exploration of this interesting phenomena (and scattered other special cases) are beyond the scope of this essay; however, a simple explanation is that in many bees, wasps, and ants, it is the number of chromosomes that determines sex: females develop from fertilized eggs, containing pairs of chromosomes; males develop from unfertilized eggs, containing only one of each chromosome.

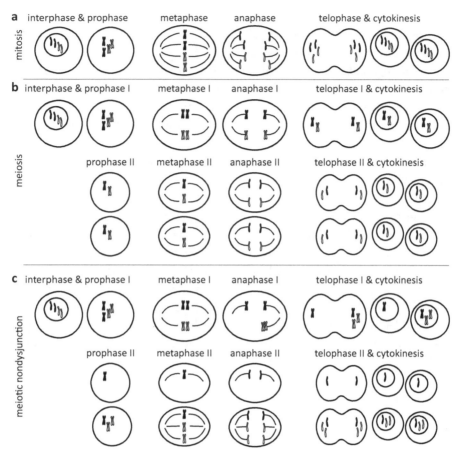

Fig. 2.4 Mitosis and meiosis. (**a**) Mitosis starts by cells moving from interphase (part of the standard 'life cycle' of a cell) to prophase: it is in interphase that the cell's DNA is duplicated, such that in prophase there are two copies of every chromosome. In prophase the membrane surrounding the nucleus (the organelle where the DNA is kept) breaks down. During metaphase the doubled chromosomes line up along the cell's centre-line; during anaphase they migrate to opposite ends of the cell. In the final phases, telophase and cytokinesis, the cell completes its division and packages the chromosomes back into a nucleus. (**b**) Meiosis echoes the stages of mitosis, running through the process twice. During metaphase I, chromosome pairs line up together (shown here as filled and unfilled), and then in anaphase I, they separate such that one chromosome of each pair migrates to opposite ends. There is sometimes a very brief interphase-like period between the first and second cycle, but no further duplication of DNA. Metaphase II and anaphase II are much like mitosis, with single copies of each doubled chromosome ending up in their new cells. (**c**) Meiotic nondisjunction occurs when an error leads to both of one chromosome pair migrating to the same side of the cell in anaphase II (here, the unfilled chromosomes). The remainder of meiosis continues as normal, resulting in four abnormal gametes: two missing the chromosome, and two with an extra copy

important differences. In *metaphase I*, the chromosomes also line up in the centre of the cell, except this time each chromosome lines up alongside its pair. In *anaphase I*, instead of the two copies of a single chromosome going to opposite ends of the cell, both copies of one of each chromosome pair go to one side and both copies of the other in the pair go to the opposite side. At this point, the cell divides. Now, there are two cells, each containing duplicated DNA—exact copies—of only one of each chromosome. The following *metaphase II* and *anaphase II* are nearly identical to metaphase and anaphase in mitosis: the duplicated chromosomes line up in the centre, then travel to opposite ends of the cell. Finally, the cells divide again. Thus, from an original germ cell containing two copies of each chromosome, we get four gametes, each of which contains only one copy of each chromosome (Fig. 2.4b).

Human DNA is organised into twenty-three pairs of chromosomes, for a typical total of forty-six chromosomes: twenty-two pairs are known as the *autosomes* (numbered 1 through 22), and the final pair consists of the sex chromosomes X and Y. Typically, males will have an X and Y chromosome, and females two X's. Male gametes (sperm) can consist of individual copies of either the 22 autosomes plus an X, or the 22 autosomes plus a Y. Female gametes (eggs) consist of the 22 autosomes plus an X. Carolyn, however, has forty-seven chromosomes: the twenty-two pairs plus three X's. An extra chromosome like this is known as a *trisomy*; Carolyn has trisomy X. Most instances of extra chromosomes create so much genetic trouble early in development that an embryo dies well before it is born. However, some human trisomies survive to birth, the most well-known of which is trisomy 21, or Down Syndrome. Trisomy X is nearly as common trisomy 21, but considerably less well-known, mainly because it often has little to no effect. Many women can live their entire lives unaware that they have an extra chromosome, as Carolyn was unaware until age sixteen.[8,9]

So, where do trisomies come from? As Carolyn explained to Susan, trisomies are due to a process called *meiotic nondisjunction*. This happens way back in anaphase I: instead of one half of each pair of chromosomes going into the new cells, one cell gets both chromosomes in the pair (and the other none).

---

[8] While many women with trisomy X have so few issues they are never diagnosed, it can also present with abnormalities including developmental delays, physical issues such as scoliosis, psychiatric disorders, and ovarian failure. One common feature of trisomy X is tall stature, and readers will note that Carolyn is physically tall (although as we have not met her parents, it is unclear if this is due to trisomy X or standard inheritance!).

[9] M Otter, CT Schrander-Stumpel & LM Curfs. 2009. Triple X syndrome: a review of the literature. European Journal of Human Genetics 18:265–271, doi: https://doi.org/10.1038/ejhg.2009.109.

The remainder of meiosis continues as normal, except at the end, two of the gametes will be entirely missing that chromosome and two will have two copies (Fig. 2.4c). Thus, when Carolyn's high school karyotype exercise revealed her three X chromosomes, everyone assumed that the third X was from meiotic nondisjunction in the formation of her mother's egg. As Susan said, the inference was that Carolyn's trisomy X "must have come from a real egg"; it was circumstantial evidence only—which in the end turned out to be incorrect!

We do not learn exactly why Carolyn's parents synthesised her with three X chromosomes. Perhaps they were prescient of the potential dangers of hiding their secret inside her DNA and did it in order to further obscure the location of their code. Had she been known to be synthesised, she could have been a target for anyone looking for material left by her parents—but not as a regular, biological offspring. Or, as Mike at one point suggests, perhaps they just needed more writing space for their message.

The Code: Synthetic Biology and Synthetic Genomes

We're getting closer to being able to explain the detail of Carolyn's code and what Carolyn's parents would have had to consider in rewriting her genome. First, though, we'll get some background about modern traditional genetic engineering and what is called *synthetic biology*, the next generation of genetic manipulation techniques, which encompasses efforts at generating synthetic genomes.

The Rise of Genetic Engineering

Development of the first main technology essential to genetic engineering predates our knowledge of DNA. In 1928, Frederick Griffith reported an experiment[10] whose design and outcome is still examined by countless biology students every year. It was the first piece of evidence along the path to identifying DNA as the genetic material. Griffith was studying the bacteria that caused pneumonia, and had noticed different appearance of colonies: those he

---

[10] F Griffith. 1928. The significance of Pneumococcal types. The Journal of Hygiene 27:113–159, doi: https://doi.org/10.1017/S0022172400031879.

called 'rough' and 'smooth'. Smooth colonies produced fatal pneumonia in mice; rough colonies resulted in no disease. In his landmark experiment, he showed that while living rough bacteria and dead smooth bacteria both left mice healthy, a mixture of the two killed the mice. Further, living smooth bacteria could be recovered from the dead mice. Thus, something from the dead smooth bacteria had 'transformed' the harmless rough bacteria into the fatal smooth type.

It was not for another sixteen years that this 'transforming agent' was identified as DNA.[11] What Griffith didn't know, and we know today, is that the smooth colonies resulted from bacteria that have an extra exterior surface known as a capsule, which protects them from the immune system. Mixing the dead smooth cells with the living rough cells resulted in the living cells taking up DNA from the smooth cells, including the DNA that coded for proteins necessary to make the protective capsule. Griffith termed the phenomena he observed *transformation* of the bacteria, and this is the term still used today to refer to the process of adding novel DNA to a cell.[12] Transformation is accomplished with a variety of techniques, all of which influence the ability of DNA to cross a cell membrane to the inside of a cell. At the opening of our story, Carolyn's PhD student Frank has been running transformations of yeast cells: he has been inserting specific sequences of DNA into his yeast.

Already having this basic ability to insert DNA into a cell, the era of genetic engineering began with the ability to produce what is known as *recombinant DNA*, which is DNA that combines sequences from different sources. This was due to *restriction enzymes*, the discovery of which Carolyn thinks about (along with green fluorescent protein) at the entrance to the C.A.V.E. Restriction enzymes are used in nature by bacteria to cut up the DNA of invading viruses, but they have been co-opted by scientists because of a key feature: most restriction enzymes cut DNA at a very specific sequence known as a *restriction site*. Restriction sites are only about four to eight base pairs long: being so short, they are highly likely to be present somewhere in a long stretch of DNA. This enables bacteria to chop up viral DNA, as it is sure to have some

---

[11] OT Avery, CM MacLeod & M McCarty. 1944. Studies on the chemical nature of the substance inducing transformation of Pneumococcal types: induction of transformation by a desoxyribonucleic acid fraction isolated from Pneumococcus type III. Journal of Experimental Medicine, 79:137–158, doi: https://doi.org/10.1084/jem.79.2.137.

[12] Transformation is used to refer to the addition of 'plain' DNA to a cell; when the DNA is packaged into a virus (as is done for some eukaryotic cells, particularly those originating in multi-cellular organisms), the process is called *transfection*.

restriction sites, and also enables scientists to piece together DNA from differ-ent sources, as the pieces of DNA a scientist wants to combine are sure to have restriction sites nearby.

When a piece of DNA is cut with a restriction enzyme, it leaves behind two *sticky ends*: short sections of single-stranded DNA, which have been separated from each other. The same restriction enzyme always leaves behind the same sticky ends, meaning that any piece of DNA cut with that restriction enzyme can bind via complementary base pairing of its sticky end to any other piece of DNA cut with the same restriction enzyme. This is how scientists can use it to combine DNA from different sources (Fig. 2.5). Much biotechnology has similar origin in an evolved, useful function in some organism. For exam-ple, PCR, or **polymerase chain reaction**, uses DNA polymerase from organ-

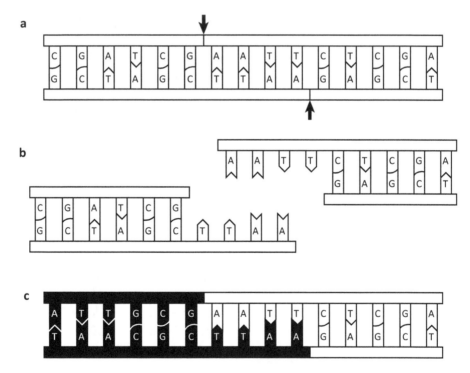

Fig. 2.5 Recombinant DNA via restriction enzymes. **(a)** DNA is cut with a restriction enzyme that recognises a particular short sequence of DNA; the arrows here show example cut sites for the enzyme EcoRI, which recognises the site GAATTC and cuts between the G and the A on both strands (the bottom strand is read right-to-left). **(b)** The DNA comes apart and a sticky end is left on each strand (with the sequence AATT). **(c)** If DNA from another source, shown here in black, is cut with the same enzyme, it has the same sticky ends. The two different pieces of DNA can bind via complementary base pairing at these sticky ends, resulting in recombinant DNA

isms that live at high temperatures, enabling scientists to repeatedly copy and melt apart strands of DNA in the lab, generating large amounts of the same DNA sequence. CRISPR/Cas9 gene editing is another technology based upon antiviral defences of bacteria, where bacteria store DNA sequences of viruses in order to recognise and destroy it; scientists use this ability to recognise a specific DNA sequence for targeting modifications to a sequence of their choice. The 'faster artificial methods' available to generate synthetic chromosomes by Carolyn's parents' time probably also co-opt natural enzymes for use in the laboratory.

As we make use of evolved biology for biotechnology, biotechnology is also moving towards design of biology for technology. Synthetic biology epitomises this idea: using biological parts as pieces in the construction of biological devices. Synthetic biology encompasses research all the way from generation of discrete functional units of DNA flanked by standardised restriction sites, enabling mix-and-match interchangeability of parts, to generation of organisms with entirely synthesised genomes.

Synthetic Genomes

The first organism with an entirely chemically synthesized genome was *Mycoplasma mycoides JCVI-syn1.0*, also known as "Synthia", in 2010.[13] Fifteen years and forty million dollars in the making,[14] Synthia was the brain-child of J. Craig Venter (founder of the J. Craig Venter Institute, source of the JCVI in the species name). The bacterial genus *Mycoplasma* was targeted for the first synthetic genome due to its small genomic size; the smallest number of genes of any independently living natural organism is found in *Mycoplasma genitalium*, with a genome only 582,970 base pairs long containing only 485 genes. This genome was artificially synthesised and propagated inside of yeast,[15] but the related *Mycoplasma mycoides* was chosen as the first synthesised genome to control its own organism due to its faster growth rate. The synthesised *Mycoplasma mycoides* genome was transplanted into another *Mycoplasma*,

---

[13] DG Gibson, JI Glass, C Lartigue, et al. 2010. Creation of a bacterial cell controlled by a chemically synthesized genome. Science 329:52–56, doi: https://doi.org/10.1126/science.1190719.

[14] RD Sleator. 2014. The story of *Mycoplasma mycoides JCVI-syn1.0*. Bioengineered Bugs 1:231–232, doi: https://doi.org/10.4161/bbug.1.4.12465.

[15] DG Gibson, GA Benders, C Andrews-Pfannkoch, et al. 2008. Complete chemical synthesis, assembly, and cloning of a *Mycoplasma genitalium* genome. Science 319:1215–1220, doi: https://doi.org/10.1126/science.1151721.

Mycoplasma capricolum, and the host genome destroyed, leaving the new cell being controlled entirely by the synthesised genome.

To generate Synthia, a modified sequence of *Mycoplasma mycoides*'s genome of slightly over one million base pairs was first synthesised in approximately thousand base pair lengths (1 Kbp fragments[16]) using a machine. The synthesised pieces were designed with some key differences from the natural living organism, including sequences to enable replication of the DNA in yeast (an intermediate stage; see below), deletions of some non-essential genes, and 'watermarks' using a DNA-based code to spell out things like the names of all forty-six authors contributing to the project and a set of famous quotations.

The 1 Kbp fragments were combined, using natural processes available in yeast: yeast contains enzymes that stitch together pieces of DNA if they have the same sequence. Thus, the small fragments were edged with lengths of overlapping sequence. Sets of ten 1 Kbp fragments were transformed into yeast at the same time, and the yeast was left to process them into the combined 10 Kbp lengths. This had an element of chance to it: overall, at least 10% of the transformed yeast cells would have the desired 10 Kbp sequence. This process was repeated to turn ten 10 Kbp sequences into 100 Kbp sequences. Eleven 100 Kbp sequences represented the entire genome, and these were again combined the same way in yeast. At this point, the yeast contained an entire, circular piece of DNA that was equivalent to a *Mycoplasma mycoides* genome, with the synthetically-added extras.

The final stage was to get the genome into its new cell. The DNA was first extracted from yeast and isolated, such that only the new genome was present. Then it was transformed into the *Mycoplasma capricolum* bacteria: while the techniques were basically scaled-up versions of standard bacterial transformation techniques, because it was an entire genome, the process was termed *transplantation*. At this point the *Mycoplasma capricolum* had two genomes: its own, plus the new, synthetic one. A key feature was that the strain of *Mycoplasma capricolum* used was broken with respect to its restriction enzyme system: the strain did not have any restriction enzymes (remember, evolved to protect bacteria from invading viral DNA), and importantly, its own genome was also not protected from being cut up by restriction enzymes. When the synthesized *Mycoplasma mycoides JCVI-syn1.0* genome joined the cell, it did code for restriction enzymes, which attacked and destroyed the native *Mycoplasma capricolum* genome as if it were invading DNA. This left the bac-

---

[16] Just like computer memory, base pairs use metric prefixes to indicate longer lengths, for example, 1 Kbp is 1000 base pairs, 1 Mbp is 1,000,000 base pairs, and so on.

terial cell with only one genome, the synthesized one, and Synthia was created.

This work continues: the creation of *Mycoplasma mycoides JCVI-syn3.0* was reported in 2016.[17] This version of the synthesized genome was minimized: from the original over one million base pairs in *Mycoplasma mycoides JCVI-syn1.0*, version 3.0 contained only 531 Kbp and 473 genes—smaller than the smallest natural genome. Interestingly, trying to determine the minimal set of genes needed based on biological knowledge failed: leaving only genes that had known, important function did not produce a viable cell. The minimal set was eventually determined experimentally, by using transposons to jump around in the genome disrupting different genes. Of the necessary 473 genes, nearly a third—149 genes—had unknown function, revealing the huge amount of biology still yet to be understood, even in an absolutely minimal, human-constructed genome.

Given the use of an intermediate stage in yeast during construction of the first synthesised genome for bacteria, it is unsurprising that yeast is the target for the first synthesised eukaryotic genome. Unlike Synthia, generated entirely within a private organisation, the Synthetic Yeast Genome Project, or Yeast 2.0, is highly collaborative involving multiple laboratories and both private and public funding. In 2017, Yeast 2.0 reported a final design for their genome, dubbed Sc2.0[18] (Sc for *Saccharomyces cerevisiae*, the species name of budding yeast, a laboratory workhorse; this is also the same yeast used in beer-making and sold dried in sachets for you to bake your own bread). The genome was designed with further genetic engineering in mind, for example, removing transposons (which might move about and change the genome on their own!), moving all the tRNA genes to a newly designed chromosome for easy access, and simplifying sequence by using a single stop codon in all genes. As of this writing in 2018, the project remains in progress, with six of yeast's 16 chromosomes having been generated and grown individually in yeast—and a subset of these combined together successfully in the same organism as well.[19]

---

[17] CA Hutchison, RY Chuang, VN Noskov, et al. 2016. Design and synthesis of a minimal bacterial genome. Science 351:aad6253, doi: https://doi.org/10.1126/science.aad6253.
[18] S Richardson, LA Mitchell, G Stracquadanio, et al. 2017. Design of a synthetic yeast genome. Science 355:1040–1044. doi: https://doi.org/10.1126/science.aaf4557.
[19] LA Mitchell, A Wang, G Stracquadanio, et al. 2017. Synthesis, debugging, and effects of synthetic chromosome consolidation: synVI and beyond. Science 355:eaaf4831, doi: https://doi.org/10.1126/science.aaf4831.

Like Synthia, the synthetic genome is built in pieces: 10 Kbp 'chunks' are designed with restriction sites on either end to match up with others via their sticky ends to build 30–60 Kbp 'megachunks'. The megachunks can then be transformed into yeast, which uses its native proteins to stitch the DNA in the right place, replacing that section of the natural chromosome. You may notice that Synthia started with 1 Kbp pieces, whereas Sc2.0 starts straight with 10 Kbp pieces: a decade on, chemical DNA synthesis has advanced such that 10 Kbp are now possible to synthesise reliably. It is not unreasonable to imagine that an entire chromosome may be possible to synthesise someday, as has happened before Carolyn's parents' time: she reflects that her parents had not needed to use yeast, since not only could the entire chromosome be synthesised in DNA, the proteins holding its three-dimensional structure could be added by machine as well.

If they did not need to, why did Carolyn's parents use yeast? First, we now know that her parents were not only working on synthetic genomes! Their work on longevity on a cellular level is well-suited to yeast, which is a eukaryotic cell (so has relevance to humans) but still a single-celled organism (meaning it can be grown quickly, and designed strains can also conveniently be frozen away and thawed out when needed—you can't do that with mice!). But for synthetic genomes, Carolyn was a bit oblivious to the obvious—perhaps due to her deliberate avoidance of anything remotely related to her parents' research area. The first synthetic eukaryotic chromosomes are currently being built in yeast due to its convenience: already one of the most well-understood single-celled eukaryotes, it is excellent for keeping and growing long sections of DNA. When a scientist wants to 'get' a bit of DNA they can either synthesise it from a machine or use PCR to copy it from nature; however, once you have it, it is far more convenient to store it in a living organism which will make as much of it as you want, whenever you want. Particularly an organism like yeast or bacteria, that you can freeze, then thaw out to grow more when you want more DNA. We've had DNA synthesis capability for decades, but the first thing a scientist does when ordered DNA shows up in the post is to transform it into bacteria for safekeeping. Similarly, Carolyn's parents would have been keeping her synthesised chromosomes safe in yeast until they were ready to put them into a human cell.

Despite the fact that we currently have yeast with partially synthesised genomes, the distance between what we can do now and what Carolyn's parents did (and then, decades later, other scientists in Carolyn's timeline) is quite large. Current research building Human Artificial Chromosomes

(HACs)[20] only provides lengths of DNA a few percent the size of a true human chromosome. The main advance enabling HACs is the fact that the cell will recognise centromeric DNA sequence and build a centromere in the right place—meaning the HAC can be transmitted alongside native chromosomes when the cell divides—but we still have no idea how this occurs. Major steps to yet cover would be generation of chromosome-sized pieces, edging them with telomeres, and packaging a full genome into a nucleus. Perhaps the closest current technology would be nuclear transfer, which transfers naturally occurring nuclei from one cell to another. This is used to make cloned organisms and 'three parent babies' (inserting the nucleus of a zygote—made from one sperm and one egg—into a second woman's egg which would have her mitochondrial DNA, typically to avoid passing on mitochondrial disease present in the first woman).

Given that clones were "old tech" by the time of Carolyn's birth, society had clearly dealt with the technical—and ethical—issues presented by nuclear transfer of the nucleus of an adult human into an egg. On the technical side, this would involve greater understanding of aging already, as it is widely believed that current issues with cloned organisms relate to aging-related signals present in the nucleus of an adult organism. One presumes that dealing with these issues also assisted in enabling ethical creation of cloned humans, removing one major barrier in the concern over the long-term health of such individuals. However, it is quite reasonable that—just as the J. Craig Venter Institute was able to create Synthia version 3.0 without understanding what one third of its essential genes actually *do*—robustly successful organismal cloning was available without complete understanding of cellular aging. Thus Carolyn's parents, in investigating how to build, not just copy, a genome, would have been well placed to stumble across essential information about aging and longevity, which they then explored in more depth.

Carolyn's Code

This longevity result is what they encoded into Carolyn's genome—the code of which we are finally placed to understand! Because they not only synthesised but also modified her genome, they actually had to cover considerably more than the advances outlined above. As Carolyn noticed, the simplest

---

[20] D Moralli & ZL Monaco. 2015. Developing *de novo* human artificial chromosomes in embryonic stem cells using HSV-1 amplicon technology. Chromosome Research, 23:105–110, doi: https://doi.org/10.1007/s10577-014-9456-2.

'proof' of her synthetic origin they could have left in her genome would have been modifying the codon code by swapping out like for like: changing both the codon usage and the frequency of tRNA genes in the genome in parallel, such that, statistically, there would be no difference in how a given mRNA would interact with the tRNAs floating about in the cell during the process of translation. As Carolyn explained to Mike, this *codon usage* is important because the more copies of the tRNA for a given codon there are, the faster that codon can be 'read' and the faster the protein can be made. Since Carolyn's parents did not swap out like for like, and instead used the codons for their own code, they had to make some additional modifications. They did this firstly in the tRNA genes: these were modified, and would have been done to make Carolyn's codon usage as close as possible to the natural. But because of the code, it could not be perfect. Thus, to fix any genes that were translated too fast or too slow, they then had to modify the promoters of these genes. This is why Carolyn's genome match was different from her parents in the coding sequences (the code) and then also the tRNA genes and a few select promoters (to fix problems generated by the code).

Before explaining the code itself, we'll answer one final question: why did Carolyn's parents use the 1.5% of her genome that is the coding sequences, rather than the other 98.5%, in which to insert their message? The main part of the answer lies in the imperfect nature of biological knowledge: because they knew what the coding sequences did. We know from ENCODE that a very large proportion of this other sequence is 'biochemically active', and as speculated above, the future society in which Carolyn's parents lived probably still did not know what most of it does. Modifying introns, or the vast amount of repetitive sequences, would be delving into these areas as-yet unknown; but they could change codons and then fix any issues using tRNA frequency and promoter modifications with confidence. Here is more science-fictional advances: we don't currently have the knowledge to tweak promoter sequences to the fine extent necessary as Carolyn's parents did, but this is within the realm of potential possibility. An additional consideration for Carolyn's parents would be that inserting the code into the transposon-related repetitive sequence that makes up the bulk of our genome runs the risk of transposons moving about and disrupting the message! Thus, it was the coding sequences only that they modified.

As Mike said, the code itself was a simple substitution, based on 'codons of codons'—sets of two to five codons could be read off as letters or numbers. As we see in Fig. 2.2, the degenerate code has variable numbers of codons for different signals: the twenty amino acids and stop. The amino acids coded for by degenerate codons can be sorted into four sets, those with two, three, four,

a

number codons	amino acids (or signal)
2	asparagine, aspartic acid, cysteine, glutamic acid, glutamine, histidine, lysine, phenylalanine, tyrosine
3	isoleucine, STOP
4	alanine, glycine, proline, threonine, valine
6	arginine, leucine, serine

b second codon

first codon

	1st	2nd	3rd	4th	5th	6th
1st	E	A	N	C	Y	7
2nd	T	I	R	U	0	9
3rd	O	S	H	F	2	V
4th	L	D	M	P	4	X
5th	G	B	1	3	5	J
6th	6	8	W	K	Q	Z

c second codon

first codon / second / third

third codon

		1st	2nd	3rd	
1st	Num	T or 1	N or 5	1st
	E or 0	O or 4	C	2nd
	I or 3	I or 9	G	3rd
2nd	A or 2	R or 7	M	1st
	S or 6	U	B	2nd
	D	Y	X	3rd
3rd	H or 8	P	K	1st
	F	V	J	2nd
	W	Q	Z	3rd

Fig. 2.6 Carolyn's code. (a) Degeneracy in the DNA codons comes with either two, three, four or six codons referring to the same amino acid or signal. (b) An example decoding based on two-codon codes for those amino acids referred to by six degenerate codons. The ordinal place in codon-usage frequency of the codon becomes the new 'character' for the code; two such characters are used to define a letter of the alphabet or a digit. The example here makes some guesses as to what Carolyn's parents may have done. For example, to maintain Carolyn's codon usage as close to natural as normal, it makes sense to encode more common letters with more common codons, thus E (the most common letter) is encoded by the combination of both first-most-common codons. The order shown here was developed based on letter frequencies in scientific writing [Found at http://letterfrequency.org/ (Accessed 5 July 2018)] with numbers stuck in near the end. (c) An example three-codon encoding using amino acids with three degenerate codons each; one set (the combination of all three most common codons) is used as a flag to indicate that the following set will encode a number instead of a letter

or six codons (Fig. 2.6a; also there are two amino acids with just one codon). What Carolyn's parents did was take each of these sets of codons and order them based on the generic human codon usage frequency: most common, second most common, and so on (this would be same frequency found in both of their genomes—what Mike was using when he cracked the code). Then, codons could be made by sets of these ordinal places. For example, the most common codon for arginine followed by the least common codon for leucine would make a codon code of: first-sixth. In our example decoding in

Fig. 2.6b, this would be read as the number 7. The genome had four overlapping sets of these codes, one for each group of degenerate codons, starting with the most codons and moving to the least.

The number of distinct elements that can be encoded by any codon-based code is based on the number of 'characters' that can be used (let's call this c: in DNA, this is the four bases; in Carolyn's code, this is the ordinal places of the codons) and the number combined into a codon (n): it is calculated as the number of characters raised to the power of the size of the codon, c^n. To understand why, take a look at the tables in Figs. 2.2 and 2.6b,c. All codons have c choices for their first character; for each of these first characters, there are again c choices for the second ($c*c$): these form the rows and columns of the tables. This process continues for each further character: each place in the codon has c separate choices, multiplying the number of possibilities again by c. Thus, for a codon of length n, the number of possibilities will be c multiplied together n times, or c^n.

Carolyn's parents used their code to encode the twenty-six letters of the alphabet plus the ten digits, so thirty-six total elements. The first pass through, using the amino acids with six degenerate codons, fits this perfectly with a two-codon code: $6^2 = 36$ (Fig. 2.6b). The remaining passes are not so perfect. Mike comments on how the final pass, those with two degenerate codons, used only a five-codon code, meaning at $2^5 = 32$ there were four missing elements—his surmise, that letters would do double-duty is quite sensible, for example O for 0, L for 1, S for 5, and E for 3. But why did they not just use six, for $2^6 = 64$? This would leave twenty-eight 'extra' codons, but we already know from DNA that you can just make a degenerate code. However, it also uses up more of your 'characters', and they would want to make the most efficient use of Carolyn's amino acids as possible. So, as it could be readable (especially at the very end, where a pattern has been set up), doubling up on a few elements makes more sense.

What about the others, those with three or four codons each? Going with their assumed efficiency, the codes on these could use a trick as shown in Fig. 2.6c for the amino acids with three degenerate codons: using three-codon codes results in $3^3 = 27$ elements. This gets us twenty-six letters of alphabet, but what about the numbers? The one remaining element could code for a 'flag' which would then indicate that the following codon should be read differently: rather than a letter, it should be read as a digit. As long as the number of digits in the whole message is not too large, because these are now effectively six-codon codes, it should be more efficient than using four-codon codes with over half degenerate elements ($3^4 = 81$). In fact, to be most absolutely efficient, rather than a flag for digits, it would make even more sense to be a flag for the least-frequent ten elements they were trying to encode. But as these were a pair of geneticists, not cryptographers, we might surmise they went with the more intuitive option.

The four-codon amino acids have exactly the same issues as DNA in trying to encode the twenty amino acids: $4^2 = 16$ is too few, but $4^3 = 64$ leaves a lot of degeneracy. The same trick as above is possible, splitting the alphabet and having flags, one for less common letters and one for digits, but this would mean a considerable number of elements actually having eight-codon codes! There is high likelihood this would be more inefficient. We'll leave it to the reader to consider whether you think Dr and Dr Schwarz would have repeated their trick with flags or lived with the degeneracy.

The Two Axes of Aging

The Schwarz Final Findings, encoded as above into Carolyn's genome, turned out to be genetic manipulations to generate longevity. This is, of course, one of the most science-fictional elements to the story. However, the features mentioned build upon current knowledge of the science of aging. We'll finish off this essay looking briefly at the biology behind the Schwarz Final Findings.

Mitochondria and Free Radicals

Carolyn directed her research into mitochondria, in the (mistaken) belief that it would keep her nice and far away from her parents' research. But, as we mentioned above, one of the current roadblocks in organismal cloning—nuclear transfer from adult cells to egg cells—is our understanding of aging's effects on DNA and the nucleus. For synthetic genomes this would have to be understood, in order to generate in the synthetic genome all the appropriate signals expected to be found in an egg. Thus, it was mostly likely in their exploration of this problem that the Schwarzes stumbled across the secret to longevity. Carolyn can be forgiven for missing this connection; by her parents' time, the problem of cloning had been solved (both scientific and societal—the latter of which was probably the less straight-forward!), and thus the problem of aging at least as it relates to nuclear transfer, and by Carolyn's adulthood, any additional understanding needed to build a genome from scratch was also known. We can surmise geniuses like the Schwarzes, in between these two stages, wandered further into aging than strictly necessary, taking them far from their synthetic chromosomes and into mitochondria.

Mitochondria are one of the organelles (remember, eukaryotes have organelles and prokaryotes do not). Mitochondria are a very special kind of organelle, known as *double-membrane bound organelles*. It is believed that these organelles have their origin as prokaryotic cells which developed an *endosymbiotic* relationship with another cell, meaning they lived together, one inside the other: the inner membrane is the membrane belonging to the prokaryotic symbiont and the outer membrane that from the eukaryotic host. As such, mitochondria have their own genome—a singular circular chromosome, just like prokaryotic cells (this is the DNA that the mitocyls analysed). The mitochondrial genome consists of thirty-seven genes, which encode ribosomal RNAs, tRNAs, and a subset of proteins needed for mitochondrial function.

Mitochondria's main role in the cell is energy production. They do this via a set of chemical reactions known as *cellular respiration*. These reactions ultimately turn chemical energy from sugars into energy stored in molecules of *ATP*, or **adenosine triphosphate**.[21] ATP forms the energy source for other chemical reactions in a cell (in plants, ATP is produced from sunlight via photosynthesis in chloroplasts, another double-membrane bound organelle); everything else a cell does can ultimately be traced back to reactions that use the energy stored in the chemical bond between the second and third phosphates in ATP, turning ATP into ADP (*adenosine **diphosphate***). Respiration is a complex set of reactions, but the components we are interested in consist of what is known as the *electron transport chain*. This is the biology that Carolyn never quite got to explain to Susan, but the implication was that this was key to her parents' longevity findings. The electron transport chain consists of a series of proteins in the mitochondrial inner membrane that pass electrons from one to another, relieving in the process the energy stored in the electrons' energy states and storing it by converting ADP to ATP.

However, the electron transport chain can be 'leaky'—some of the electrons can escape, generating what are known as free radicals: these are extremely reactive molecules, containing an unpaired electron, which go on to react with and damage other molecules in the cell, ranging from DNA to

---

[21] The astute reader may notice the linguistic similarity between adenosine triphosphate and the base adenine mentioned at the very start of this essay. This similarity goes beyond words: adenine is a component of ATP, which additionally consists of the sugar ribose and three phosphate groups—a similar molecule to a single adenine nucleotide of RNA, except with three phosphate groups instead of one. Further related are the 'dNTPs' in the very first bit of decoded material that Carolyn read on the airplane: dATP, or *deoxyadenonsine triphosphate*, is one of the raw building blocks of DNA, along with dGTP, dTTP, and dCTP, collectively known as *deoxynucleoside triphosphates*—two of the phosphate groups are removed when a dNTP joins a growing chain of nucleotides. The 'deoxy' part simply indicates that the sugar is deoxyribose instead of ribose.

the cellular membrane. The Free Radical Theory of Aging suggests that presence of free radicals, and their resulting damage, is a large component of cellular aging. The theory was originally proposed in the 1950s, and while it has had some criticisms, current thought is that there is still strong potential for its validity.[22] Thus, the Schwarzes' concentration on the electron transport chain—the source of the leaks—as well as mitochondrial turnover—because older mitochondria have leakier electron transport chains[23]—forms one axis of their double-headed attack on aging.

Telomeres and Cellular Senescence

The other prong of the Schwarzes' longevity treatment relates to a chromosomal feature known as telomeres: these are the very ends of eukaryotic, linear chromosomes. We join Carolyn's story just as she is giving a lecture explaining the unique problem generated by keeping DNA in a linear, rather than circular, piece. When DNA is copied, it always starts out with an *RNA primer*. This is a small bit of RNA that the cell adds using complementary base pairing to get the copying process started. This is necessary because the cell cannot start building a DNA strand in isolation; the enzyme that builds a DNA molecule, known as *DNA polymerase*, can only add to a growing strand. RNA polymerase, in contrast, can start a strand of RNA with a single nucleotide (as it does when making an mRNA[24]). However, RNA and DNA are close enough that DNA polymerase can use a few nucleotides of RNA as a starting point, or 'primer', and continue on, building DNA. Thus, the process of copying DNA always begins with an RNA primer. Importantly, RNA and DNA can only be built in one direction: by adding a new nucleotide that connects its phosphate group to the sugar of the last nucleotide in the growing strand.

In a circular prokaryotic chromosome, the RNA primer starts the copying process, then, when the growing DNA strand comes back around the circle to reach it from the other side, the RNA primer is replaced with new

---

[22] LCD Pomatto & KJA Davies. 2018. Adaptive homeostasis and the free radical theory of ageing. Free Radical Biology and Medicine 124:420–430, doi: https://doi.org/10.1016/j.freeradbiomed.2018.06.016.

[23] DA Chistiakov, IA Sobenin, VV Revin, AN Orekhov & YV Bobryshev. 2014. Mitochondrial aging and age-related dysfunction of mitochondria. BioMed Research International 2014:238463. doi: https://doi.org/10.1016/j.atherosclerosis.2008.05.036.

[24] There are in fact many types of RNA polymerases and DNA polymerases; the RNA polymerase that makes mRNA is typically not the same RNA polymerase that makes an RNA primer. However, all RNA polymerases and DNA polymerases share the features described here: RNA polymerases can start with a single nucleotide, DNA polymerases must add to a strand.

DNA. However, in a linear eukaryotic chromosome, there will always be one RNA primer, the *terminal RNA primer* (a term Carolyn uses in that first lecture), that has no DNA on the other side. Because DNA only grows in one direction, the piece of DNA opposite the RNA primer is never copied. This happens every time a linear piece of DNA is copied, meaning it constantly gets just a bit shorter.

As Carolyn was explaining, chromosome ends consist of telomeres, which are the same DNA sequence repeated thousands of times (this would be part of the large component of the genome that is repetitive sequence). It is these repeats that get slowly eaten away as cells divide and generate new copies of their DNA for the new cells. They form a buffer that keeps the important functional sequences (like genes) in the interior of the chromosome safe. Carolyn ended her first lecture pointing out that this cannot go on forever: no matter how many repeats are present in the telomere, if it gets shorter each time a cell divides, it would eventually vanish. Her second lecture covered the solution: an enzyme known as telomerase. This enzyme is active in the germ cells which give rise to gametes (in the process of meiosis, covered above), and its function is to grow back the telomeres. In this way, each new generation starts with full-length telomeres: a restored buffer around the central, information-containing DNA sequence.

However, in the rest of the body, telomeres do continue to shorten. Is thought that these shortening telomeres serve an important function: they signal to the body cells that are older and thus more likely to have accumulated DNA and other types of damage. Cells only divide fifty to seventy times before entering what is known as *senescence*, a halting of the grow-and-divide cycle. In some parts of the body, old cells will undergo *apoptosis*, programmed cell death. In this way, the body has evolved to protect itself from rogue cells with damaged DNA. Many of the genes known as tumour-suppressor genes, genes that protect us from cancer, are involved in senescence and apoptosis. Cancer cells often have abnormally short telomeres, reflecting their continued dividing, as well as that they have gotten around the protections against such uncontrolled cell growth provided by senescence and apoptosis.

Telomeres also serve another function for linear chromosomes: they signal to the cell that the end of the chromosome is not a mistake to be fixed. Otherwise, DNA repair processes in the cell would try to patch together the 'broken' end of DNA with other chromosomal DNA, potentially causing severe genetic problems in that cell. As telomeres shorten, it is possible this signalling is reduced, leading to age-related DNA damage. Additionally, the three-dimensional structure of telomeric regions is thought to have regulatory influences on gene expression in the rest of the chromosome, which may be modified as telomeres shorten. All of these features contribute to the current belief that telomere length serves an important, although as-yet fairly unknown, contribution to cellular aging.[25] We can presume that the Schwarzes' research did reveal the mechanisms behind this, and thus their focus on telomeres and the enzymes involved in their regulation (like telomerases) formed the second axis of their longevity treatment.

A Longevity Treatment

The Schwarz Final Findings combined what Carolyn's parents had found about mitochondria and telomeres to generate a longevity treatment. In this final section of our essay, we'll speculate on what form this treatment may have taken. They synthesised Carolyn's genome with the telomeric modifications already in place, but we know that they tested the treatment on themselves. Thus, it would need to be genetic modifications that could be performed on adult organisms. This modification would need to address both the genomic DNA—and its telomeres—and the mitochondrial DNA—including the proteins in the electron transport chain.

We are closer to understanding how the genomic DNA may have been modified. There are a number of mechanisms for modifying the DNA of mammalian cells, what is currently termed 'gene therapy' in humans. Until recently, the best method of modifying DNA in an adult mammal would be through processes similar to transformation of bacteria: the addition of genetic material to a cell. This can be done via specifically designed viruses that do not cause disease, but infect target cells and deliver a 'payload' of the desired DNA. Some of these viruses can cause the DNA to integrate into the chromosomal DNA of the infected cell, meaning that the cell would carry the new DNA through its divisions. Defective proteins could either be 'knocked down' by production of RNA that interfered with its mRNA, or via corrective DNA repair via presentation of inserted correct sequences.

However, the new technology of CRISPR/Cas9 gene editing, briefly mentioned above, has much greater potential for being a useful genetic modification tool. Rather than just adding DNA to a cell, it enables changing the sequence, in place, on the chromosome. CRISPR stands for **C**lustered **R**egularly **I**nterspaced **S**hort **P**alindromic **R**epeats, which are areas of bacterial

---

[25] G Lidzbarsky, D Gutman, H Shekhidem, L Sharvit & G Atzmon. 2018. Genomic instabilities, cellular senescence, and aging: *in vitro, in vivo* and aging-like human syndromes. Frontiers in Medicine 5:104, doi: https://doi.org/10.3389/fmed.2018.00104.

genomes that consist of the same bit of DNA repeated many times, but separated by a variety of different unique sequences. The purpose of these were a mystery, until scientists discovered that, like restriction enzymes, they served as part of protection against viruses. The unique sequences are stored pieces of viral sequence that the bacterial cell can then recognise and destroy.

The way the bacteria do this is transcribe these CRISPR sequences into RNA, which is held by *CRISPR-associated proteins* (*Cas proteins*). When the RNA binds to its complementary sequence (from a virus), the Cas protein chops it up. To use this for biotechnology, the viral sequence is replaced with the genomic sequence you want to change. The Cas protein will chop up the native genomic sequence, and then one of two things can happen. If you simply want to destroy that bit of DNA, for example to make the gene nonfunctional, you leave it be and the DNA repair processes mentioned above will fill in random sequence to fix the DNA, resulting in a broken gene. If you have a specific sequence change you want to make, you can provide a template of desired sequence for the DNA repair processes to use as a model, such that after chopping up your recognised sequence it would be replaced by the sequence from the template.

A mechanism like CRISPR/Cas9[26] may be what the Schwarzes used in their treatment, or a similar yet-to-be-discovered process. One feature that would need to be addressed, in order to provide a longevity treatment to an adult organism, would be how to modify the DNA of all cells in the organism, since non-modified cells would continue to age, or to include sequence that might enable a programme that resulted in wholesale replacement of all non-modified cells. Alternately, it may be that their longevity research identified only a few important cell types that were necessary to change, such as stem cells.

The other modifications in the longevity treatment were to the mitochondria. Mitochondria DNA modification are an area of active research, but there is no clear path as of yet suggesting what technique could be most promising. Current investigations on how to insert DNA into mitochondria use mechanisms ranging from viruses to nanoparticles, but none have been hugely successful.[27] The hypothetical Boltzer Process, discovered first by the Carolyn's parents and then later by Ethan, provides a science-fictional solution to this problem. By seeing that Carolyn's students were doing this on a regular basis,

---

[26] The '9' in Cas9 simply indicates it is the ninth in the family of Cas proteins, which has been particularly useful for biotechnology.

[27] AN Patananan, TH Wu, P-Y Chiou & MA Teitell. 2016. Modifying the mitochondrial genome. Cell Metabolism 23:785–796, doi: https://doi.org/10.1016/j.cmet.2016.04.004.

we can surmise the process by Carolyn's time was as common and easy to use as bacterial transformation is today.

Some Final Words

What's been glossed over in much of this essay, and apparently is not prominent by Carolyn's time with human cloning being passé and designer babies common, are the ethical and societal issues associated with the futuristic biotechnology described above. Biotechnology is rapidly developing, and while much of the above seems infeasibly far away, other technologies, like organismal cloning and gene editing, are already here. Less than sixty years—a time period well within modern human lifespan—separated the description of DNA's structure in 1952 from the generation of an organism controlled entirely by human-synthesised DNA in 2010. It may be nearly impossible to speculate on what we might be capable of doing in the space of another lifetime. What is clear, though, is that consideration of ethical implications and impacts on society must keep apace with developing biotechnology.